# HOW TO WIN EVERY ARGUMENT

## The Use and Abuse of Logic
### Second Edition

# MADSEN PIRIE

B L O O M S B U R Y

LONDON • NEW DELHI • NEW YORK • SYDNEY

**Bloomsbury Academic**

An imprint of Bloomsbury Publishing Plc

| 50 Bedford Square | 1385 Broadway |
| London | New York |
| WC1B 3DP | NY 10018 |
| UK | USA |

**www.bloomsbury.com**

**BLOOMSBURY and the Diana logo are registered trademarks of Bloomsbury Publishing Plc**

First published 2006, second edition first published 2015

**British Library Cataloguing-in-Publication Data**
A catalogue record for this book is available from the British Library.

ISBN: PB: 978-1-47252-912-1
ePDF: 978-1-47252-396-9
ePub: 978-1-47252-697-7

**Library of Congress Cataloging-in-Publication Data**
A catalog record for this book is available from the Library of Congress.

Typeset by Deanta Global Publishing Services, Chennai, India
Printed and bound in India

To Thomas, Samuel and Rosalind

# CONTENTS

# ACKNOWLEDGMENTS

My thanks for their helpful suggestions go to Eamonn Butler and John O'Sullivan. For assistance with the preparation, I thank Tom Lees, Steve Masty, Sam Nguyen and Xander Stephenson. I also thank all those who have aided and encouraged this work, not least the publisher and editor.

# INTRODUCTION

Sound reasoning is the basis of winning at argument. Logical fallacies undermine arguments. They are a source of enduring fascination, and have been studied for at least two-and-a-half millennia. Knowledge of them is useful, both to avoid those used inadvertently by others and even to use a few with intent to deceive. The fascination and the usefulness which they impart, however, should not be allowed to conceal the pleasure which identifying them can give.

I take a very broad view of fallacies. Any trick of logic or language which allows a statement or a claim to be passed off as something it is not has an admission card to the enclosure reserved for fallacies. Very often it is the case that what appears to be a supporting argument for a particular contention does not support it at all. Sometimes it might be a deduction drawn from evidence which does not sustain it.

Many of the fallacies are committed by people genuinely ignorant of logical reasoning, the nature of evidence, or what counts as relevant material. Others, however, might be committed by persons bent on deception. If there is insufficient force behind the argument and the evidence, fallacies can add enough weight to carry them through.

This book is intended as a practical guide for those who wish to win arguments. The section covering how to argue tells the reader how to construct arguments that will win the day and more importantly, perhaps, how to conduct arguments successfully in a variety of different contexts. It also teaches how to perpetrate fallacies with mischief at heart and malice aforethought. I have described each fallacy, given examples of it, and shown why it is fallacious. After any points of general interest concerning the history or occurrence of the fallacy, I have given the reader recommendations on how and where the fallacy may be used to deceive with maximum effect.

I have listed the fallacies alphabetically, although a full classification into the five major types of fallacy may be found at the end of the book. It is well worth the reader's trouble to learn the Latin tags wherever possible. When an opponent is accused of perpetrating something with a Latin name it sounds as if he is suffering from a rare tropical disease. It has the added effect of making the accuser seem both erudite and authoritative.

In the hands of the wrong person this is more of a weapon than a book, and it was written with that wrong person in mind. It will teach such a person how to argue effectively, even dishonestly at times. In learning how to argue, and in the process of practising and polishing each fallacy, the user will learn how to identify it and will build up an immunity to it. A working knowledge of these fallacies provides a vocabulary for talking about politicians and media commentators. Replacing the vague suspicion of double-dealing will be the identification of the precise crimes against logic which have been committed.

Knowledge of fallacies can thus provide a defensive as well as an offensive capability. Your ability to spot them coming will enable you to defend yourself against their use by others, and your own dexterity with them will enable you to be both successful and offensive, as you set about the all-important task of making arguments go your way.

Madsen Pirie

# HOW TO ARGUE

## 1 What is an argument?

An argument starts with a disagreement. The first person puts forward a point of view or an opinion, and the second person disagrees, putting across a point or opinion that is not compatible with what the first person has said. By incompatible, we mean that they cannot both be held simultaneously. It is entirely possible that one is correct and the other incorrect, or that both are incorrect. But if they are incompatible they cannot both be correct.

> *I think there is life on Mars.*
> *I maintain that Mars is lifeless.*

(They are incompatible. You can support either one or the other, but not both. Since they contradict each other, if one is true, the other must be untrue, and vice versa.)

> *I think the Duchess was wearing a black dress when she opened the new college.*
> *I think she was wearing a white dress on that occasion.*

(They, too, are also incompatible. You can maintain either of them but not both, and in this case you can opt for other alternatives. She might have been wearing red or green, for example.)

Given a disagreement, an argument represents the attempt by each side to establish the validity of the position they are putting forward, and to undermine the opposing case. They can do so by bringing forward evidence in support of the position they are maintaining, or by pointing to evidence which casts doubt on the alternative position.

Although we might loosely use the term argument when people simply assert opposing positions, it is more correctly applied when they bring forward supporting material. It is not really an argument if they simply repeat their initial assertion, no matter how often they do so. Nor is it really an argument if they simply deny the truth of what their opponent claims. In the famous sketch from *Monty Python's Flying Circus*, the man enters room 12A to have an argument:

*Ah, you're having an argument with me.*
*No I'm not.*
*Yes you are.*
*No I'm not.*
*Yes you are.*
*Etc.*

*This isn't an argument.*
*Yes it is.*
*No it isn't.*
*Yes it is.*
*Etc.*

The humour of the sketch notwithstanding, the exchange is a disagreement rather than an argument unless supporting material is produced on either side.

That supporting material should be in the form of evidence or additional argument rather than simple vehemence.

*The column that supports Nelson's statue in London's Trafalgar Square is over 185 feet high.*
*No, it's actually under 170 feet.*
*Look! You know damn well it's over 185 feet!*

(and a fist pounding the table in this exchange is not considered to add much intellectual backing to the case)

The arguments about life on Mars, the colour of the dress the Duchess wore and the height of Nelson's Column have one thing in common: they are about facts. There is or is not life on Mars, the Duchess's dress was of a particular colour and Nelson's Column is either above or below 185 feet high. These are arguments that can be resolved by supporting evidence. Respected authorities can be cited by either side to reinforce their case.

In the case of Nelson's Column, an official report following the 2006 restoration of the statue included laser measurement of its height above street level. This was put at 169 feet by 3 inches, well short of the 185 feet previously quoted in textbooks. This report describing the measurement is sufficiently authoritative that it is now generally accepted that the monument is some 16 feet less tall than previously thought. The production of this evidence settles the argument to the satisfaction of most observers.

The colour of the dress the Duchess wore when she opened the college would have been seen by people in the crowd that witnessed the event. There were probably photographs of it published in the local or national press. There may have been television coverage of the event, accessible online or in the archives of the TV station or anyone who happened to record it. Images of the dress in print or in video recordings will provide evidence to settle the question of whether the dress was black or white. When this evidence is included in the argument, it will generally prove decisive in determining its colour.

The presence of life on Mars is also a question of fact, but the difference here is that it is not (in 2014) yet a known fact. There is not yet evidence out there in the public domain that can establish the certainty or otherwise of life on Mars. Living organisms have not been photographed, televised, or subject to conclusive chemical examination that could settle the issue. Note that the two sides, while contradictory, are not symmetrical (i.e. not on an equal footing).

It would take the discovery of one living organism to settle the argument in favour of life there, but failure to find life would not prove there was none, only that none had yet been found. It is quite possible that extensive examination of large parts of the surface of Mars might show no life, but this would not prove there was none elsewhere on the planet, or even underneath its surface.

Even though the fact itself is not known, there can still be evidence advanced to support the two sides of the argument. Those who think there is life on Mars can point to the undoubted evidence that the planet once had ample surface water, a precondition of some forms of life. They can point to evidence that its atmosphere once contained a chemistry that could have supported life. In a much-publicized move, NASA scientists announced that a meteorite on Earth of Martian origin had been found to contain fossilized traces of primitive ancient life. Even though other scientists have advanced alternative explanations of the finding, it remains a piece of possible evidence.

On the other side, those who think Mars is lifeless can cite its inhospitable conditions, its extremes of temperature, its current dryness and the radiation that, in the absence of a magnetosphere or a thick atmosphere, bathes its surface. All of these suggest, they say, that conditions on Mars are not suited to life. As for the evidence on the other side, they say that it might support the case that life had once existed long ago when Mars was different, but do not support the case for life currently existing there.

The argument about life on Mars differs from that concerning the colour of the dress worn by the Duchess or the height of Nelson's Column. All three hinge on questions of fact, but in the case of the dress and the monument there is ample, readily accessible supporting evidence available. In the Mars case there is not.

There was a time until quite recently when arguments about fact would feature large in pub conversations. People would argue about many subjects, especially involving sport, and would make assertions that others would challenge.

*Higgs scored more Premier League soccer goals before Christmas than any other player has ever done.*
*Wrong. Christianson scored more for Spurs pre-Christmas in the 2004 season.*

The arguers would challenge each other's alleged facts, sometimes heatedly, sometimes scornfully. In the absence of any way of looking at supporting evidence, they would sometimes appeal to their friends, and arguments would occasionally be settled by a democratic vote among those present as to what they thought did or did not happen.

Not any more. Similar arguments or disagreements still happen in pubs, but now they take minutes to resolve instead of the lengthy and heated discussions they used to give rise to. The difference has been caused by the ubiquity of smartphones and online sources of information. Now when contentious statements are made, after a brief disagreement someone will take out a smartphone and look up the Google or Wikipedia entry on the subject, or some similar source of information, and produce it as compelling evidence to resolve the matter.

In the above hypothetical argument about Premier League soccer goals, there would be readily accessible information online to establish whether the pre-Christmas crown belonged to Higgs or Christianson, and the discussion would be over, with victory awarded to whomever the online evidence favoured.

Smartphones and access to the internet have changed arguments significantly. People had to wait until they returned home or visited a library to check which positions the available evidence supported. Now they do it on the spot. It means that there is little point any more in arguing about questions of fact. Such questions can now be resolved by instant access to authoritative evidence that will settle most of them. Disagreements that used to provoke argument now serve as a stimulus to finding out what the facts are. The internet and the smartphone give everyone on-the-spot access to the world's libraries, and the means of resolving disagreements about facts.

Some people's social life has been ruined by this development. These are the types whose idea of a pleasant evening used to be to assert a position or to ridicule one put forward by a companion, and then to browbeat and bully the other party into submission. This could involve red-faced shouting which they greatly enjoyed, and more often than not they would end the evening satisfied that they had 'won' any arguments. They might later discover that they were in the wrong, but this in no way undermined the pleasure. The point was to win on the night. Nowadays when such people try it, a smartphone is produced within minutes and the issue is settled, leaving them seething with resentment at being deprived of an evening's entertainment.

Most arguments about fact are no longer worth engaging in. They are too easily resolved, and any skill and intellectual effort devoted to persuading others of the certainty of an alleged fact is wasted if the outcome can be checked online.

There are two types of argument about fact that are not so easily resolved. One is the sort where the facts are not known, such as the possibility of life on Mars. In such cases it is worthwhile to marshal supporting evidence that might convince others.

The other type is where the facts are disputed, where there is evidence but people are not agreed on its significance. This type of argument is illustrated in many scientific disputes. We might agree on what is observed, but disagree and argue over its possible causes. Often there are two sides to such disputes, each conducting research designed to provide evidence that one particular theory should win out over its rivals.

We are reasonably sure that dinosaurs became extinct quite rapidly some 66 million years ago, but there has been debate about the causes of that event. One school of thought has argued that there must have been a cataclysmic event such as an asteroid or comet striking the Earth and devastating the habitat. Another school has suggested that volcanic activity poisoned the atmosphere. Each side has attempted to find evidence that supports their argument. The 'collision' school has revealed a thin worldwide layer of iridium laid down 66 million years ago – an element rare on Earth but more prevalent in asteroids and comets. The 'volcanoes' school has produced statistical analysis seemingly showing that the time frame of the extinction was too long to be the result of an impact, and must be put down to a longer-term cause such as an increase in vulcanism. The argument continues.

'Global warming' has proved a highly contentious argument about fact, with some disputing that it is taking place at all, and others disputing the evidence that human activity is a cause of it. It is about fact, but about disputed facts, and each side has attempted to diminish or discredit the evidence advanced by the other. The argument has been conducted with more rancour than is customary, even in hotly contested scientific disputes, and this could be because more than reputations hinge on its outcome. Ultimately it is about who shall have the ear of government and influence it to act. The opposition to global warming has been as ill-tempered as its proposition, alleging that the switch to 'climate change' represents the fallacy of definitional retreat, stepping back from 'global warming' something remarkable, unusual and potentially alarming, into 'climate change' that happens all the time. Proponents have responded by dubbing the opposition 'deniers', to link them in the public mind with those who deny the holocaust. This level of rancour and abuse is not the way that rational arguments should be conducted.

The field of historical explanation affords many examples of arguments about fact that cannot be resolved by appealing to authoritative evidence to clinch the decision one way or the other. The facts of history happened; the argument is about what caused them to happen, and whether those alleged causes were facts themselves. It often comes down to interpretation, sometimes even about whether the so-called facts happened at all.

We know the English Civil War and the French Revolution happened, but people disagree about why they happened. Some cite economic causes, showing a rise in expectations among a newly emerging middle class. Others cite political causes. Some opt for intellectual influences, citing the works of contemporary writers and a developing lack of respect for authority, whereas others look to changes in religious beliefs. Each pores away over such data as can be gleaned, advancing evidence in support of their favoured explanation.

Whereas many scientific disputes are eventually settled in favour of one theory over its rivals, historical disputes seldom are. One explanation seems to gain ascendancy and enjoy a fashionable popularity, only for it to be challenged later by newly uncovered evidence and lose its ascendancy. Different modes of interpretation such as Marxism or feminism go in and out of fashion and influence the way we look at history in their turn. Arguments over historical causes seem to drag on for decades, even centuries, without being resolved firmly in favour of one preferred explanation. This is good for the publishers of history textbooks, since they have to be replaced periodically as their accounts of events go out of fashion. But it is confusing to students of history seeking to form an account in their minds of what happened and why.

There is an important point that we would like to learn the lessons of history and to avoid repeating the mistakes of the past. If we cannot agree on what caused past events, we are poorly equipped to take measures that might prevent them from being repeated. The Great Depression of the 1930s was a terrible event that devastated the lives of millions, both in the United States and across Europe. Some see it as the catalyst that led to political extremism and ultimately World War II. Yet the argument still rages as to what caused it, and as to whether the measures taken to ameliorate it in fact made it worse. Research yields evidence that supports various different positions, and there is no general agreement as to what its causes were, much less on how to prevent them recurring.

The debate over the causes of the Great Depression shows that argument can be more than just an amusement to fill leisure time, and more than a contest to see who can come out top dog. There are occasions in which argument matters, when its outcome will determine how we act, and whether our condition will be improved or made worse by what we do. There are occasions when argument can affect our very survival.

*I say we open the door and head out now before the fire reaches us. No. If we open the door the air will rush in and make the fire spread more rapidly.*

(The warm feeling of victory in an argument is nice, but there is more at stake in this one.)

———

# 2  How a successful argument should look

A common type of successful argument starts with the present position that people agree upon. It points to the conclusion that the speaker wishes to establish. It assembles the evidence in support of that conclusion, and it deals with contrary evidence that might undermine it.

*We agree we both want to go out tonight. I suggest we make it the cinema rather than bowling.*
*There are several new movies showing that you've expressed an interest in, including that one that all our friends are talking about.*
*A movie would be over in time for us to get home in time for a pleasant nightcap before bed.*
*It's been a tiring day and we'd both enjoy a good sit-down for a couple of hours.*

*I know you like bowling, and I enjoy it, too, but we're both pretty tired and it's a pretty strenuous activity.*
*The bowling alley is farther away than the cinema, so we'd have to get the car out instead of taking public transport.*

*We can book seats online for the cinema, but we don't know how long we might have to wait for a free lane at the bowling alley. John and Enid are coming over next week, and we could go bowling then with them and make it a foursome.*

This is a very traditional form of argument. From where we are now there are two courses we could take. Here are the reasons in favour of what I recommend, and here are the objections to the alternative. It looks very similar if more than one alternative course is under consideration, but the structure is the same: the points in favour of mine are set alongside the points against the alternatives.

If this argument were to be expressed as a visual diagram, it might look like the skeleton of a fish. From the head at the top, where we start, the central column takes us in a straight line down to the conclusion to be established, the tail. On the left are the reasons or the evidence advanced in support of that progression, like the bones of a fish, and on the right are the refutations of contrary evidence or reasoning, or possible evidence against alternatives that might be proposed. By the time we have reached the conclusion we have presented evidence for our case and undermined evidence against it.

It is very important to have a clear view about what counts as evidence. It can come from the past, showing examples of how similar things have been done before and have produced results that were deemed satisfactory.

*Whenever we have built houses with private lawns, these have always been better maintained and less prone to criminal activity than have common areas shared by everyone on the estate.*

(This does not prove that private gardens will always improve the appearance of housing estates and cut crime there, but in saying that they have done so in the past, it will be taken as an argument that they might do so again in the future.)

*You always resist going on holiday to the Lake District because it rains so often, but every time we've done it before you've thoroughly enjoyed it and come back elated.*

(It might not work this time. There might come a point when enough rain is enough, but the previous enjoyment is at least an argument in favour of going again.)

The supporting evidence can come from the experience of other people, as well as that of those involved in the argument.

*You say we'd find Disney World too organized and commercial, but the fact is that every one of our friends who's been there has come back raving about it.*

(Again, the experience of others does not *prove* anything conclusively. We might be different from our friends in crucial respects, but without evidence of that difference, their experience does count in support.)

This appeal to the experience of others is especially useful when something is proposed that has been tried in other countries. Its success there can be taken as evidence that it might work here, though do expect opponents of the proposal to argue that what works in other countries will not work here because of differences in culture, history or geography.

*I say we should include geothermal energy in our list of renewable sources. It works brilliantly in Iceland.*
*Yes, but our geology is totally different from that of Iceland. We'd have to dig deeper to access it, and that would make it much more costly.*

Most people, but not all, put greater store by what has actually been done than they do by what might be done. In folk wisdom people say that 'an ounce of practice is worth a pound of theory'. Evidence in support of an argument that draws on what has happened in the past is therefore given greater weight by most people than reasoning about what might happen in the future. Most people value experience ahead of speculation.

*I read some expert saying that you can lose weight if you spend an hour a day sitting inside a glass pyramid.*
*I've heard that most people do it by eating sensibly and exercising more.*

The first is advanced as evidence of a kind, since 'some expert' has endorsed it. It is, however, about what *might* happen. 'You *can* lose weight.' The second is also evidence of a kind, since it draws on the experience of others. The difference is that it is about what they actually do, not what they might do, and therefore will seem more convincing to most people. Furthermore, the evidence presented to those encountering this argument will probably lead them to draw on their own experience. Many will know people who lost weight by dieting and exercise, but very few, if any, will know of cases where sitting inside a glass pyramid did the trick.

The question of what does or does not count as supporting evidence for an argument is a complex one. Broadly speaking, when trying to make a case or win an argument you are trying to convince other people to come round in support of what you are saying. Your supporting material should be such as to help that to happen. It should be that which convinces your audience that it is relevant to the case you are making, that it supports that case and that it is itself reliable and trustworthy.

When barristers make arguments in law courts, they deploy supporting evidence that will command the respect of judge and jury. They routinely call 'expert' witnesses, and establish the authority of the witness to pass judgement on the matter under consideration. The 'expert' will be someone with qualifications that make their considered opinion carry more weight than might that of an uninformed casual observer. They might introduce a witness with a distinguished career in forensic science to give an opinion as to what can be inferred from the medical evidence. They might call a person renowned for their work on ballistics technology to say what can reasonably be deduced from the state of the bullets or cartridges or from the wounds they are alleged to have caused.

The opinions of those with qualifications are taken to count for more than those of people who have no specialized knowledge of the subject, but to carry weight they must be qualifications in the field the discussion is about. Lobbyists and pressure groups commonly cite in support of their arguments a large number of 'scientists' who have no expertise in the field under consideration. The fact that seven nuclear physicists support organic farming does not add much support to that style of agriculture. Listeners might be more impressed if the 'scientists' had some expertise in the biological sciences instead. Similarly, the half-dozen movie stars produced in favour of changing policies on the environment or development aid do not add much to the argument, since neither their acting skills nor their ability to take instructions from a film director are relevant to the policies being advanced.

Even in cases where the 'authority' has secured some position or won some accolade in the field, it might not indicate expertise in the specific subject they are claimed to support.

> *Using more debt to escape from a recession is backed by a leading Nobel Prizewinner in Economics. That gives it powerful backing.*

> *Yes, but his prize was awarded for work done decades ago on international trade patterns and geographical wealth distribution. That doesn't qualify' him to tell us how to achieve economic growth.*

Arguments are advanced in letters to newspapers supported by multiple signatures, with the aim of supporting the argument by the authority of the names listed. Sometimes it is legitimate, with a collection of genuinely qualified people. But just as the internet enables arguments about fact to be quickly resolved, it also enables the credentials of alleged authorities to be examined. Sometimes we find that the group of businessmen supporting some recommended trade policy include those with no great merit in business experience, let alone trade policy. And the large numbers of 'economists' hauled out in favour of proposed tax changes seem to include retired accountants and lecturers in business studies from little-known US community colleges. The moral is that if you want to bring in the support of 'experts' on your side, make sure they are credible experts.

Supporting evidence used to make an argument more convincing has to be relevant to what is being considered. The experiences of female workers in a sewing factory might not bear all that strongly on the safety conditions appropriate for shipyard workers or those engaged in steelmaking. If the relevance of your supporting evidence is not immediately obvious to your audience, you should explain what does make it relevant to your argument.

———

# 3  Oral arguments

Oral arguments take place in a variety of settings and situations, and the style of argument varies accordingly. It is still about advancing a position, providing evidence that supports it and defending it against opposition, but the manner of doing so can be varied to make it more appropriate to the circumstances. People would think it strange if someone having an argument in a pub with a friend started declaiming in the style of a public debate before a mass audience. This is because context matters.

## With a friend

When you argue in private with a friend it is well to keep your temper and to keep your voice down. You are more likely to appear reasonable if you put your case forward in level tones. State what you have to say without shouting. By all means be passionate if the subject is something you care about. Indeed, your strength of feeling, if you let it show, can demonstrate your commitment and help reinforce your case. But remember that an argument is not the same as a row. In a row people often abuse each other for things done or left undone, but an argument is supposed to be an appeal to reason, an attempt to persuade, and no amount of fist-banging and shouting at the top of your voice will help. They are more likely to lose you friends rather than convince them.

Be courteous. After you make your point, allow your friend to make his or hers. A reasoned discussion is less likely to take place if those arguing constantly interrupt each other. Television presenters interrupt their subjects before they have answered the point, but they are not arguing, just trying to show their superiority in front of their audience to boost their ratings. Friends should listen to each other and try to counter the points that each of them makes.

Since your aim is to win the argument with your friend, the evidence you produce in favour of it should, where possible, be calculated to appeal to what you know about that friend. If they are widely travelled and not given to xenophobia, for example, you might draw on the experiences of people in other countries to add to your case. If they take a keen interest in sport, or practise it themselves, analogies and examples from the world of sport might persuade them. The aim is to select evidence likely to resonate with their own experience.

Not only listen to your friend, but show you are listening by acknowledging their point and then explaining why you do not find it convincing. Your key phrases to begin your response are 'Yes, but . . .' and 'That's all very well, but . . .'. Then you should explain why special circumstances mean that in this case their point does not undermine your argument.

## With a group

Arguing among a group of people in a pub or coffee bar requires different techniques. The group will not share the same interests, so examples to win one over might have no appeal to the others. And some people will introduce side issues that threaten to switch the argument to something of more interest to themselves. Keep it relevant, sometimes explaining politely that what they are saying might well be true, but is not really relevant to the discussion at hand. You should focus on the case you are making and keep returning to it.

In a group discussion there will often be people on your side, and it helps hugely if you cite their agreement and reinforce something they have said that backs your case. If you bring in the occasional 'As Peter said . . .', it makes it clear that you are not alone in thinking what you do.

There will often be a main opponent in a group discussion, someone who makes most of the running against your case. Rather than allow it to settle into a two-person argument where the others are mere onlookers, you should try to draw them in by asking some of them what they think about certain aspects of the matter. Most people like to be asked their opinion, and it provides the chance of bringing in extra support and evidence.

Most important in a group discussion is to control the course of the argument. You know the point you want to put across, and you want the discussion to centre around the evidence you put forward to advance that point. The key here is analysis. You should, if you can, list the main areas you think bear on the case at hand, and start the discussion on them and try to keep it there.

*'I think there are four main points to consider here. First . . .'*

Breaking down the subject like this into specific areas is a powerful way to direct the agenda of the discussion. People might disagree with some of your points, but they will talk about them instead of the other things they could have brought up that might be less supportive of your position.

## A formal public debate

A formal public debate lacks the cut and thrust, the immediacy of point and counter-point that can characterize arguments with a friend or in a group. A public debate has its rules, and these normally give each speaker a limited time in which to make their points. A typical format sees two speakers on each side of a motion, speaking in succession. Quite often audience members might make brief points from 'the floor' before a speaker sums up on each side at the end.

This format means that instead of answering the opposition's points as they are made, speakers have to try to deal with them all at once in their own speech afterwards. Going first gives you the chance to direct the course of the argument, and going last gives you the chance to have the last word without being contradicted.

The first rule about winning a public debate is preparation. You do not need to prepare your whole speech, but you should at least list on a card the points you intend to make in support of your side of the argument. If you need to, you can write out the whole speech, but try to deliver it, rather than reading it. Delivery is important, and you should try to make your argument sound spoken, rather than written, remembering that the two have somewhat different grammars. If you can, speak spontaneously from the notes you have made; it sounds more convincing than reading from a prepared text. It has the added advantage that you can make eye contact with your audience as you speak, rather than staring down, with your eyes glued to your text.

It helps your argument if you can, in addition to making notes of all the points you wish to bring up on your side, anticipate the things the other side might say and undermine them in advance. It looks weak and unconvincing if they then stand up and make a point you have already refuted. If you are speaking after an opponent, it helps to make brief notes of their points as they are speaking, so you can refute some of them when your turn comes.

The second rule is to get the audience on your side if you can. If you can have them warm to you, they are more likely to give attention to your side of the argument. A little humour, especially at the outset, helps the audience to relax, and engages their attention.

The third rule is to be systematic. You should list, one by one, the supporting evidence you are giving to put forward your case. This can add a remorselessness to your side of the argument as you produce one after another the reasons why your side should prevail.

The crucial difference between a public debate and an argument between a friend or friends is that you are not trying to convince your opponents. Instead of trying to win them round, as you do with friends, you are trying to persuade third party observers, the audience, that you are right and that your opponents are wrong. This means that the courtesies are different. In an informal argument with friends it is usually counterproductive to mock or insult them, but in a public debate you can do both of these things to your opponent. A certain degree of civility is expected, but you can pour scorn on them in a way you would not do to people you are trying to win over. To see this difference writ large, watch the House of Commons arguments at Prime Minister's Question Time. Insults, scorn and mockery are the order of the day, and can be very effective if done with humour and panache.

There should be a definite conclusion to the speech. It looks rather tame if a speaker finishes his argument and then sits down. If you have time, it is often useful to briefly remind the audience of the points you have made, and to point out that they add up to a powerful case for your side of the argument. You should finish with a flourish that serves as an applause line when you sit down.

*'It is for all these reasons, Mr Chairman, Ladies and Gentlemen, that I call upon you to reject this utterly misguided motion and vote with the opposition.'*

(Or to support this completely worthwhile and sensible motion and vote in its favour)

## Arguing before a public meeting

There are many occasions when people are called upon to present a case before an audience. It might be that Jones is required to present a case to the Board in support of a proposed acquisition. Col. Travers might have to address the Parish Council on why the proposed leisure centre development should be stopped. The circumstances are often those that involve proposing or opposing an initiative. Speakers frequently find that their argument has to be against the status quo if they are proposing, and in defence of it if they are opposing.

In both cases the structure of their argument should be like the fish skeleton referred to earlier. They should start with the present situation, tell the audience where they are headed, then list the points in favour of their case on one side and undermine the points against it on the other. The supporting arguments should be ones that will resonate with the audience by appealing to their outlook and interests.

If, for example, the case is against a proposed development, it should point to the damage it will do to the local environment, to the distress and nuisance it will cause the local community and to the dislocation and stress it will involve over the many months or years it will take to complete it. On the other side of that fish skeleton, the points might cast doubt on the proposal's alleged benefits, disputing the number of jobs its advocates claim it will bring, doubting the various claims made for the convenience it will bring, and suggesting that it will exceed the cost that has been claimed for it, and exceed the published time it will take to complete it. Most of the audience will be aware of similar projects elsewhere for which some or all of these things have been true, and that knowledge reinforces the argument when it is appealed to.

As with a public debate, it is good advice to prepare notes that list the points that support the argument, and to round off the case with an emphatic conclusion. One difference is that an argument presented to a public meeting is not usually against another speaker, but is often for or against a proposed initiative. It is an attempt to convince an audience to do something or to refrain from doing something. It should weigh the advantages and disadvantages of that action in the pans of a balance scale, augmenting one and diminishing the other, and tilting it heavily down in favour or against the action, depending on which side the speaker wants to win.

# 4  Arguing in the media

Arguments in the media, be it newspapers and magazines, online social media or radio and television, tend to be briefer. This is because most media space is crowded, with contributions being limited in order to make space and time for other material. A premium is placed on brevity (especially on Twitter), and someone who can argue cogently and succinctly finds it easier to gain media slots than those who are long-winded and take time to get to the point.

## Arguing in newspapers

For most people access to newspapers is via the letter columns. Those who know the trade might persuade editors to run short articles from time to time, but for the non-professional Joe Public, it means the letters page. Some letters editors are kind people who will take the time to edit long letters down to usable size, but most are not. The moral is: keep it short.

Your argument should relate to a news item or article that has appeared in the paper. Your letter might start 'Concerning your report that . . .'. Then comes your argument. You can make two or three snappy points in response, and you can win the argument either by reinforcing or destroying the original report.

> *'You reported Councillor Jones' claim that the new by-pass was "good value" because it brought shorter travel times and reduced accidents. Yet the Council's own survey shows that it brought more vehicles onto the roads and that the average time saved is only 1.5 minutes per journey. Furthermore, although there has been a slight reduction in minor collisions that cause vehicle damage, police figures show no reduction in accidents involving personal injury. Given that the money spent on the by-pass could have built ten schools, Councillor Jones obviously has a strange sense of value.'*

If you want to convince a journal's readership, it helps to have some idea in your mind of who those readers might be. Will they be business people on their morning commute, or maybe tradesmen and women or shop assistants giving it a quick read over breakfast? Tailor your arguments so they will be calculated to appeal to the type of people who'll be reading it.

# Arguing on the internet

Internet arguments often take place in the comment responses to blogs. Some regular internet commentators attract large numbers of readers, and many newspapers now feature regular on-line columnists.

People take issue with what is written and reply with a comment. Others respond to that, and the argument thread can involve several people and go on for days. Some points are worth noting. Not many people who read blogs bother to go through the comments section below them, and those who do are probably not representative of the typical reader. They are an interested minority. It is also true that many, maybe most, of those who read comments are themselves quite opinionated, and difficult to sway. Many of them have minds already made up and will not change them. This means that the comments are really a form of jousting. The aim is not to convince people, but to score points off those who hold differing opinions.

It is also true that the level of courtesy is much lower than would be expected in face-to-face confrontation. Perhaps the impersonal nature of the internet makes people ruder than they would be to anyone physically present. Internet trolls hide behind comparative anonymity to engage in vilification that they might never do in person. Abuse is common, and people feel quite free to insult the intelligence and even the moral character of their opponents in ways that would be counterproductive in face-to-face arguments and would risk losing the goodwill of most audiences.

Keep it short. Of the minority who read comments, few will scroll down to the next screen, so the arguments most likely to be read are those which are compressed to a single screen. This might suggest a maximum of perhaps 360 words, depending on the blog and the computer.

Make it factual. The one thing that works in a blog comment argument is a good fact on your side or one against your opponents. Google will point you in the direction of facts that militate against the opposition's case, and quoting them will earn you battle points.

Be immediate. A rapid response that comes while the subject is hot is better than one coming two days later when it has cooled. A rapid exchange of comments conjures up the atmosphere of a ding-dong exchange that might happen when opponents are facing each other.

## Winning an argument in 140 characters

Twitter is faster still. Its brevity gives it an immediacy. To argue on Twitter, people have to be ready to exchange rapid-fire responses. Discussions move on quickly, and it is difficult to make a point and command attention when the discussion has moved on.

Twitter is also very popular, and its threads on hot topics can be followed by millions. The numbers who follow comments on internet blogs is small by comparison. Twitter requires fairly instant responses, and many stories are dead after an hour or two as discussion moves on and new topics engage attention.

The 140-character limit means you have to be very brief indeed. You cannot really hope to make more than one point in a single tweet. If you have more to say, say it in another tweet, or in a series.

*'She wants to spend my money on the poor, on the unemployed, & on free meals to children of rich parents. How does that make her generous?'*

This can be followed up in fairly short order by another, associated point.

*'She spends her own money on her villa in Tuscany, on flying there & back, and on expensive school fees for her children.'*

(This is a good argument-winner because most people have a down on hypocrisy.)

You can add force to your side of the case by retweeting to your own followers and good points made on your side of the case by others.

When someone tweets simple abuse, you should rapidly tweet back that abuse is what people who lack arguments resort to. This is more effective than counter-abuse, though an immediate witty comeback will usually attract approval.

# Arguing in a radio or TV discussion

Unless you are a panelist on a discussion show such as 'Question Time' or 'Any Questions,' your presence will probably be as someone calling in on a phone-in show, or as someone interviewed by a presenter to comment on a current news item. As with other media, time is at a premium and you must present your argument in a concise and coherent manner. What producers and presenters value is a sound-bite, which is a sentence or two that encapsulates a point of view succinctly.

Presenters will often ask a series of questions, and what they value most is someone who can answer each in a time-limited sound-bite. It used to be that a 40-second answer was what they thought ideal, but with today's shorter attention span many of them look for 25-second answers. You can say a fair amount in 40 seconds and manage to work in two of three pieces of evidence that support your position.

Preparation is helpful. If before the interview you jot down a few facts that support your argument, you will be better able to remember them when the question comes up. Any statistics have to be kept simple in an oral argument. When an argument is made in writing, people can pause, look it over, consider and absorb it. In an oral discussion they do not have that option because the speaker has moved on.

*'Polls show that 47.8% of all young people aspire to owning and running their own business one day.'*

(This is fine in print, but on radio or TV make it 'nearly half of all young people'.)

It helps, if you can, to think ahead even while you are making one point as to how you might take on the next one and how you might finish your answer. Some presenters feel they have to put on an air of hostility with people they interview, and put questions that make it difficult for their subjects to develop their argument. The key here is to find phrasing that will move the discussion back to what you came to talk about and to the points you intended to make. Watch how seasoned politicians do it, and pick up a few of their techniques.

> *A*: *'Isn't it true that we now have a higher level of unemployment than when you were in opposition?'*
>
> *B*: *'What is true is that we have more people in work than ever before. Only last month 2,000 new jobs were created in construction, 5,000 in retailing and 8,000 in services.'*
>
> (Note how the answer did not address the question directly, but did enable the supporting facts to be brought in.)

On a phone-in show, it is worth noting that most callers simply give their opinions. The ones who give facts, supporting evidence and arguments are therefore at an advantage in bringing others round to agreeing with them. The moral is: don't just say what you think; say why you think it.

―――――

# 6 Going beyond the logic

To win an argument is to come out top in a verbal duel. You are attempting to convince an opponent to switch to your point of view, or taking on an opponent in front of an audience and attempting to win over that audience. You want them to take your side.

It is possible to win an argument despite behaving in a very unpleasant way. You can be bad-mannered, objectionable and generally offensive, yet still win if the force of your reasoning cannot be withstood. That said, it is easier to win an argument if your demeanor is pleasant and if either your opposition or your audience warm to you. Salesmen know this instinctively. There is an old saying to the effect that you always sell yourself as well as the product, and it also applies to argument.

Nobody likes to lose, although some can handle defeat more graciously then others. Former world chess champion Bobby Fisher said he liked to win at chess 'to crush his opponent's ego'. If you make that your objective in argument you will meet with more determined resistance. A skilled arguer will sometimes offer an opponent a let-out, a graceful retreat that allows them to concede without losing face. It can be presented as something that looks like a compromise, allowing them some element of right in their position.

*'I think we are in agreement about what needs to be done, and the difference is only about the most effective method. I think what I've been suggesting might be a more effective way of achieving the goals we both share, so the difference between us is quite minimal.'*

You can do this in front of an audience, emphasizing the areas of agreement between you and your opponent, making it look as though you are asking them to support you on a point of detail, rather than inflicting a crushing defeat on your opponent.

Your behaviour towards your opponent matters because it colours the audience's perception of you. Courtesy is best, and good manners will stand you in better stead than churlish behaviour. The occasional wickedly barbed and sly comment will win far more support than sustained abuse.

In winning support it is important to appear reasonable rather than dogmatic. Audiences tend to show reserve towards glib speakers who seem to think they know all the answers. A little hesitancy goes a long way.

*'I find myself thinking that, despite finding much merit in my opponent's position, the arguments persuade me on balance that the truth lies elsewhere. Let me share some of those arguments with you.'*

In chess there is a well-known stratagem of sacrifice, in which a player will allow an opponent to capture a pawn or minor piece in order to gain a major piece or a winning position. The metaphor derived from this speaks of 'giving up a pawn to gain a queen', and it extends beyond chess. In argument it might involve conceding trivial points in order to appear reasonable and undogmatic, while giving up none of the ground that matters.

*'My opponent rightly points out that for some people homeopathic remedies lead to a cure, and he is right to point to significant numbers who think homeopathy is successful in treating a variety of conditions. I also concede that it does indeed have a loyal band of supporters. I am simply saying that its scientific basis is not sufficiently established for us to pay for its inclusion with NHS treatment.'*

This seems much more reasonable than saying, 'Homeopathy is unscientific and should not be part of the health service.'

It does no harm to praise an opponent for their eloquence, humour or charm, for everything, indeed, except the quality of their argument.

*'For Brutus is an honourable man*
*So are they all, all honourable men . . .'*

(And remember that Antony won that argument.)

Two further things go a long way in establishing rapport with an audience. One is a smile and the other is eye contact. Everyone except Ebenezer Scrooge likes a smile, and most people warm to those who do it. Eye contact is a way a speaker can establish individual liaison with members of the audience and make them feel that he or she is talking to them. The speaker who looks up to the ceiling or addresses the back wall of the room will not find members of the audience giving them the same attention as if they felt themselves spoken to directly.

When arguing on a blog, it is better to phrase arguments as if you were addressing one person. If you avoid saying things like 'you all' or anything that implies you are addressing a crowd, the individual sitting looking at a screen can feel you are speaking directly to them. Similarly on television, look directly into the camera lens as if it were a person's eyes, and every viewer will feel you are making eye contact with them. Even if there is a monitor showing another speaker or a presenter, ignore it and stare into the lens as if it were a face.

None of this strictly speaking adds to the intellectual quality of your argument. You do that by marshalling your facts and supporting evidence and by exposing the errors and weaknesses in your opponent's position. But bear in mind that you are trying to win over an audience, and they are more likely to be sympathetic to your point of view if they regard you as a reasonable and decent person. They will pay more attention to what you say, and will probably give you the benefit of any doubt.

In putting forward an argument you are appealing to reason, and will be more successful if you yourself appear to be reasonable. And the best way to seem reasonable is to actually be reasonable. Conduct your argument in a level and civilized manner. Just be sure that you win.

# Abusive analogy

The fallacy of abusive analogy is a highly specialized version of the *ad hominem* argument. Instead of the arguer being insulted directly, an analogy is drawn which is calculated to bring him into scorn or disrepute. The opponent or his behaviour is compared with something which will elicit an unfavourable response toward him from the audience.

*Smith has proposed we should go on a sailing holiday, though he knows as much about ships as an Armenian bandleader does.*

(Perhaps you do not need to know all that much for a sailing holiday. Smith can always learn. The point here is that the comparison is deliberately drawn to make him look ridiculous. There may even be several Armenian bandleaders who are highly competent seamen.)

The analogy may even be a valid one, from the point of view of the comparison being made. This makes it more effective, but no less fallacious, since the purpose is to introduce additional, unargued, material to influence a judgement.

*If science admits no certainties, then a scientist has no more certain knowledge of the universe than does a Hottentot running through the bush.*

(This is true, but is intended as abuse so that the hearer will be more sympathetic to the possibility of certain knowledge.)

The fallacy is a subtle one because it relies on the associations which the audience make from the picture presented. Its perpetrator need not say anything which is untrue; he can rely on the associations made by the hearer to fill in the abuse. The abusive analogy is a fallacy because it relies on this extraneous material to influence the argument.

*In congratulating my colleague on his new job, let me point out that
he has no more experience of it than a snivelling boy has on his first
day at school.*

(Again, true. But look who's doing the snivelling.)

While politicians delight in both abuse and analogies, there are
surprisingly few good uses of the abusive analogy from that domain. A
good one should have an element of truth in its comparison, and invite
abuse by its other associations. All other things being equal, it is easier
to be offensive by making a comparison which is untrue, than to be
clever by using elements of truth. Few have reached the memorable
heights of Daniel O'Connell's description of Sir Robert Peel:

*. . . a smile like the silver plate on a coffin.*

(True, it has a superficial sparkle, but it invites us to think of
something rather cold behind it.)

The venom-loaded pens of literary and dramatic critics are much more
promising springs from which abusive analogies can trickle forth.

*He moved nervously about the stage, like a virgin awaiting the
Sultan.*

(And died after the first night.)

Abusive analogies take composition. If you go forth without
preparation, you will find yourself drawing from a well-used stock
of comparisons which no longer have the freshness to conjure up
vivid images. Describing your opponents as being like 'straightlaced
schoolmistresses' or 'sleazy strip-club owners' will not lift you above
the common herd. A carefully composed piece of abusive comparison,
on the other hand, can pour ridicule on the best-presented case you
could find: 'a speech like a Texas longhorn; a point here, a point there,
but a whole lot of bull in between'.

# Accent

The fallacy of accent depends for its effectiveness on the fact that the meaning of statements can change, depending on the stress put on the words. The accenting of certain words or phrases can give a meaning quite different from that intended, and can add implications which are not part of the literal meaning:

*Light your cigarette.*

(Without accent it looks like a simple instruction or invitation.)

*Light your* cigarette.

(Rather than the tablecloth, or whatever else you feel in the mood to burn.)

*Light* your *cigarette.*

(Instead of everyone else's.)

Light *your cigarette.*

(Instead of sticking it in your ear.)

Even with so simple a phrase, a changed accent can give a markedly changed meaning.

*We read that men are* born *equal, but that is no reason for giving them all an equal vote.*

(Actually, we probably read that men are born equal. *Born* equal carries an implication that they do not remain equal for long.)

Accent is obviously a verbal fallacy, for the most part. Emphasis in print is usually given by italics, and those who supply them to a quotation from someone else are supposed to say so. In speech, however, unauthorized accents intrude more readily, bringing unauthorized implications in their wake. The fallacy lies with the additional implications introduced by emphasis. They form no part of the statement accepted, and have been brought in surreptitiously without supporting argument.

The fallacy of accent is often used to make a prohibition more permissive. By stressing the thing to be excluded, it implies that other things are admissible.

> *Mother said we shouldn't throw* stones *at the windows.*
> *It's all right for us to use these lumps of metal.*
>
> (And mother, who resolved never to lay a *hand* on them, might well respond with a kick.)

In many traditional stories the intrepid hero wins through to glory by using the fallacy of accent to find a loophole in some ancient curse or injunction. Perseus knew that anyone who *looked* at the Medusa would be turned to stone. Even villains use it: Samson was blinded by the king of the Philistines who had promised not to *touch* him.

Your most widespread use of the fallacy of accent can be to discredit opponents by quoting them with an emphasis they never intended. ('He said he would never *lie* to the American people. You will notice all of the things that left him free to do.') Richelieu needed six lines by the most honest man in order to find something on which to hang him; with skilful use of the fallacy of accent you can usually get this down to half a line.

It is particularly useful when you are advocating a course of action which normally meets with general disapproval. Accent can enable you to plead that your proposed action is more admissible. ('I know we are pledged not to engage in germ warfare against people in *faraway* lands, but the Irish are not far away.')

When trying to draw up rules and regulations, bear it in mind that there are skilled practitioners of the fallacy of accent quite prepared to drive a coach and six through your intentions. You will then end up with something as tightly worded as the old mail monopoly, which actually spelled out that people shouting across the street could be construed as a breach of the mail monopoly. (They did only say the *street,* though.)

# Accident

The fallacy of accident supposes that the freak features of an exceptional case are enough to justify rejection of a general rule. The features in question may be 'accidental', having no bearing on the matter under contention, and may easily be identified as an unusual and allowable exception.

> *We should reject the idea that it is just to repay what is owed. Supposing a man lends you weapons, and then goes insane? Surely it cannot be just to put weapons into the hands of a madman?*

(This fallacy, used by Plato, lies in not recognizing that the insanity is an 'accident', in that it is a freak circumstance unrelated to the central topic, and readily admitted to be a special case.)

Almost every generalization could be objected to on the grounds that one could think of 'accidental' cases it did not cover. Most of the general statements about the consequences which follow upon certain actions could be overturned on the grounds that they did not cover the case of a meteorite striking the perpetrator before the consequences had occurred. To maintain this would be to commit the fallacy of accident.

It is a fallacy to treat a general statement as if it were an unqualified universal, admitting no exceptions. To do so is to invest it with a significance and a rigour which it was never intended to bear. Most of our generalizations carry an implicit qualification that they apply, all other things being equal. If other things are not equal, such as the presence of insanity or a meteorite, the exceptions can be allowed without overturning the general claim.

> *'You say you have never met this spy. Can you be sure he was never near you in a football crowd, for example?'*
> *'Well, no.'*
> *'When was this occasion, and what papers passed between you?'*

(If I did meet him, it was an accident.)

Accident is a fallacy encountered by those in pursuit of universals. If you are trying to establish watertight definitions of things like 'truth', justice' and 'meaning', you must not be surprised if others spend as much energy trying to leak the odd accident through your seals.

Plato was searching for justice. John Stuart Mill, trying to justify liberty except where there is harm, or serious risk of harm, to others, found himself forever meeting objections which began, 'But what about the case where . . .?' It is an occupational hazard. If you are to avoid accidents, avoid universals.

> *Promises should not always be kept. Suppose you were stranded on a desert island with an Austrian count who was running an international spy-ring. And suppose there was only enough food for one, and you promised him . . .*

(The only amazing feature of these lurid stories is that anyone should suppose such freak cases to make the general rule any less acceptable.)

One of the famous examples of the fallacy is a schoolboy joke:

> *What you bought yesterday you eat today. You bought raw meat yesterday, so you eat raw meat today.*

(With the generalization referring to the substance, regardless of its 'accidental' condition.)

The fallacy of accident is a good one for anarchists because it appears to overturn general rules. When it is claimed that you are breaking the rules, dig up the freakiest case your imagination will allow. If the rule does not apply in this case, why should it apply in yours? ('We all agree that it would be right to burn down a tax office if this were the only way to release widows and orphans trapped in the cellar. So what I did was not *inherently* wrong . . .')

# Affirming the consequent

To those who confuse hopelessly the order of horses and carts, *affirming the consequent* is a fallacy which comes naturally. An occupational hazard of those who engage in conditional arguments, this particular fallacy fails to recognize that there is more than one way of killing a cat.

> *When cats are bitten by rabid hedgehogs they die. Here is a dead cat, so obviously there is a rabid hedgehog about.*

> (Before locking up your cats, reflect that the deceased feline might have been electrocuted, garrotted, disembowelled, or run over. It is *possible* that a rabid hedgehog got him, but we cannot deduce it as a fact.)

The arguer has mixed up the antecedents and consequents. In an 'if . . . then' construction, the 'if' part is the antecedent, and the 'then' part is the consequent. It is all right to affirm the antecedent in order to prove the consequent, but not vice versa.

> *If I drop an egg, it breaks. I dropped the egg, so it broke.*

> (This is perfectly valid. It is an argument called the *modus ponens* which we probably use every day of our lives. Compare it with the following version.)

*If I drop an egg, it breaks. This egg is broken, so I must have dropped it.*

(This is the fallacy of affirming the consequent. There could be many other incidents leading to a broken egg, including something falling upon it, someone else dropping it, or a chicken coming out of it.)

For valid logic we must affirm the first part in order to deduce the second. In the fallacy we affirm the second part in an attempt to deduce the first. Affirming the consequent is fallacious because an event can be produced by different causes. Seeing the event, we cannot be certain that only one particular cause was involved.

*If the Chinese wanted peace, they would favour cultural and sporting exchanges. Since they do support these exchanges, we know they want peace.*

(Maybe. This conclusion might be the most plausible, but there could be other, more ominous reasons for their support of international exchanges. The cat can be killed in more ways than one.)

This fallacy receives a plentiful airing in our law-courts, since it is the basis of circumstantial evidence. Where we have no eyewitness evidence, we work back from what is known to those actions which might have caused it.

*If he had been planning murder, he would have taken out extra insurance on his wife. He did take out extra insurance.*

*If he intended poison, he would have bought some. He did buy some weedkiller.*

*If he had wanted to cut up the body, he would have needed a big saw. Such a saw was found in his toolshed.*

(There could be alternative explanations, innocent ones, for all of these actions. It would be fallacious to say that any of them proved him guilty. But as they mount up, it becomes progressively easier for twelve good persons and true to eliminate reasonable doubts about coincidence. No doubt they are sometimes wrong and thereby has hanged many a tale, together with the occasional innocent man.)

This is an extremely good fallacy to use when you wish to impute base motives to someone. Motives do not show, but the actions caused by motives do. You can always gain a hearing for your suggestion of less-than-honourable motives, by use of a skilfully affirmed consequent.

*She's just a tramp. Girls like that always flaunt themselves before men, and she did appear at the office party wearing a dress that was practically transparent!*

(We can all see through this one.)

# Amphiboly

Amphiboly is the fallacy of ambiguous construction. It occurs whenever the whole meaning of a statement can be taken in more than one way, and is usually the fault of careless grammar.

*The Duchess has a fine ship, but she has barnacles on her bottom.*

(This is a duchess who requires especially careful handling.)

The fallacy is capable of infinite variation. Many excellent examples of amphiboly make use of the confused pronoun: does the 'she' refer to the ship or to the Duchess? Similar confusion may occur with animals.

*I met the ambassador riding his horse. He was snorting and steaming, so I gave him a lump of sugar.*

(Would that all diplomats were so cheaply entertained.)

Misuse of the word 'which', or its omission for brevity, both produce many classic examples. ('On the claim form I have filled in details about the injury to my spine which I now enclose.') There are innumerable versions of the advertisement:

FOR SALE: Car by elderly lady with new body and spare tyre.

The mistake usually consists in the failure to appreciate that an alternative reading is possible. Sometimes the punctuation is misplaced; sometimes there is not enough of it to eliminate the ambiguity. Press headlines, with their need for both punch and brevity, are favourite long grasses from which the occasional delightful amphiboly will bounce into view. Legendary World War II masterpieces include:

MACARTHUR FLIES BACK TO FRONT

(With more variations still if the second word is taken to be a noun.)

FRENCH PUSH BOTTLES UP GERMANS!

(Hand-to-hand combat, yes. But this is ridiculous.)

Use of the amphiboly with intent to deceive is a favourite resort of oracles and fortune-tellers. A timely amphiboly enables the prophet to hedge his bets, having it both ways. After the outcome one can always take refuge in the meaning which *was* fulfilled. Croesus asked the oracle what would happen if he attacked Persia. The reply 'A mighty empire will be humbled' was prophetic indeed. But it was his own.

To become a skilled perpetrator of amphibolies you must acquire a certain nonchalance toward punctuation, especially commas. You must learn to toss off lines such as 'I heard cathedral bells tripping through the alleyways', as if it mattered not a whit whether you or the bells were doing the tripping. You should acquire a vocabulary of nouns which can be verbs and a grammatical style which easily accommodates misplaced pronouns and confusions over subject and predicate. The astrology columns in popular newspapers provide excellent source material.

# The analogical fallacy

The analogical fallacy consists of supposing that things which are similar in one respect must be similar in others. It draws a comparison on the basis of what is known, and proceeds to assume that the unknown parts must also be similar.

*The body politic, like any other body, works best when there is a clear brain directing it. This is why authoritarian governments are more efficient.*

(None of these false analogies likening the state to a human body ever seem to say much about its liver, pancreas, or waste-disposal mechanism.)

Analogies are a useful way of conveying information. They enable us to talk about the new concept in terms which the audience already have experience of. The fallacy comes in the assumption of further similarities in the future on the basis of the ones already identified.

*Babies are like the weather in their unpredictability.*

(They are also wet and full of wind.)

It is fallacious because analogies are tools of communication more than sources of knowledge. An analogy may suggest lines of enquiry to us, but it does not provide a basis for establishing discoveries.

*She had skin like a million dollars.*

(Green and crinkly?)

Analogical fallacies abound in the interpretation of history. In the attempt to make history mean something, all kinds of comparisons emerge. Past civilizations all have it in common that they are now past, once were civilizations, and before that were not. These three utterly commonplace facts lead many historians into a 'life-cycle' analogy. The simple sequence 'not alive, alive, no longer alive' irresistibly invites comparison with living organisms. Before our defences are ready, there we are with civilizations 'blooming' and 'flowering', soon to be engaged in the act of 'withering and dying'.

*As our culture ripens, it is only natural that it should, like any organism, put out seeds to reproduce itself in distant places.*

(An argument for colonialism which should be nipped in the bud.)

The fact is that civilizations are not flowers. If you fall into the analogical trap, you will soon be having them drawing strength from the soil, and perhaps even exhibiting their blooms in turn.

Hume, in his *Dialogues Concerning Natural Religion,* has the earnest Cleanthes compare the universe to a delicate mechanism, like a watch. And, just as we can deduce from a watch the necessary existence of a watchmaker, so from the universe . . . But the sceptical Philo kills the argument at this point by saying that the universe seems to him much more like a cabbage.

The analogical fallacy is devastatingly effective when used against the person who first produced the analogy. Everyone uses analogies of sorts; all you have to do is seize upon one used by an opponent and continue it in a way more conducive to your own line of argument. With luck, your opponent will be forced into the admission that his own analogy was not a very good one and will lose points with the audience.

*'As we sail forth on our new committee, may I express the hope that we can all pull together for a smooth voyage.'*
*'The chairman is right. But remember that rowers were usually put in chains and whipped. And if the ship sank, they went down with it.'*

You will go far in any organization by likening it to a family. Family life evokes a pleasant glow, and the analogy will enable you in practice to argue for almost anything, including giving pocket money to the members and sending the naughty ones supperless to bed.

# Anecdotal argument

An anecdote is a particular story, whereas a general assertion tells us what happens in general. An opposing anecdote does not tell us that the general assertion is wrong, only that something different happened in a particular case. By providing a counter-example, it does indeed prove that the generalization is not a universal, but few of them are claimed to be such. The generalization tells what generally happens, not what always happens, and a contrary anecdote does not disprove it.

*A: We have tried to make the welfare system encourage people into work wherever possible because we think they are better off that way.*

*B: Well I have a letter here from a lady in Newcastle who says that her son-in-law attempted suicide when his benefits were withdrawn. That's not exactly better off, is it?*

The point is that it might be true that people in general might be better off if encouraged into work, and an anecdote, or even a handful, of cases where this did not happen do not really refute the generalization. Basically statistics should be countered by statistics, rather than by individual cases. To counter an argument of principle with a few contrary cases is to enter the fallacy of anecdotal reasoning. In honour of a popular English commentator, it might be called the *argumentum owenjonesium*.

When you come up against a generalization that looks plausible and might even be true, it is usually possible to dig up some obscure case in which it didn't apply. Even without proving opponents wrong, you will have made their case look weaker in the eyes of the audience. At the very least your opponent will now have to argue why your anecdote is not a typical one, and be diverted from the original thrust of their argument.

> *'It's all very well to say that chocolates and crisps make you fat, but I met three people only yesterday for whom this was simply not true.'*

(With luck you won't be asked to name them.)

# *Antiquitam, argumentum ad*

Students of political philosophy recognize in the *argumentum ad antiquitam* the central core of the arguments of Edmund Burke. Put at its simplest, it is the fallacy of supposing that something is good or right simply because it is old.

> *This is the way it's always been done, and this is the way we'll continue to do it.*

(It brought poverty and misery before, and it will do so again . . .)

There is nothing in the age of a belief or an assertion which alone makes it right. At its simplest, the *ad antiquitam* is a habit which economizes on thought. It shows the way in which things are done, with no need for difficult decision-making. At its most elevated, it is a philosophy.

Previous generations did it this way and they survived; so will we. The fallacy is embellished by talk of continuity and our contemplation of the familiar.

While the age of a belief attests to experience, it does not attest to its truth. To equate older with better is to venture into the territory of the fallacy. After all, human progress is made by *replacing* the older with the better. Sometimes men do things in a particular way, or hold particular beliefs, for thousands of years. This does not make it right, any more than it makes it wrong.

*You are not having a car. I never had a car, my father never had one, and nor did his father before him.*

(Which is probably why none of them got anywhere.)

The Conservative Party is the home of the *ad antiquitam.* They raised it and by golly they are going to keep it. The old values must be the right ones. Patriotism, national greatness, discipline – you name it. If it's old, it must be good.

The commercial world is sensitive to the prevalence of the fallacy, and modifies its actions accordingly. A cigarette brand called Woodbine, with a large market share, feared its image was becoming dated, but did not wish to shatter the instinctive preference for the traditional. A science fiction magazine called *Astounding* feared that its name reflected an earlier era and might hold back its development. In both cases the decision was made to effect gradual change, with the cigarette-packet design and the magazine name both changing imperceptibly over the weeks. *Astounding* made it into *Analog,* but Woodbines seem to have disappeared without trace. Perhaps cigarette customers are more conservative than science-fiction readers?

Skilful use of the *ad antiquitam* requires a detailed knowledge of China. The reason is simple. Chinese civilization has gone on for so long, and has covered so many different provinces, that almost everything has been tried at one time or another. Your knowledge will enable you to point out that what you are advocating has a respectable antiquity in the Shin Shan province, and there it brought peace, tranquillity of mind and fulfilment for centuries.

*We make our furniture in the best way; the old way.*

(And it's every bit as uncomfortable as it always was.)

# Apriorism

Normally we allow facts to be the test of our principles. When we see what the facts are, we can retain or modify our principles. To start out with principles from the first (*a priori*) and to use them as the basis for accepting or rejecting facts is to do it the wrong way round. It is to commit the fallacy of apriorism.

*We don't need to look through your telescope, Mr Galileo. We know that there cannot be more than seven heavenly bodies.*

(This was a short-sighted view.)

The relationship between facts and principles is a complicated one. We need some kind of principle, otherwise nothing presents itself as a fact in the first place. The fallacy consists of giving too much primacy to principles, and in not permitting them to be modified by what we observe. It makes an unwarranted presumption in favour of a theory unsupported by the evidence, and therefore rejects evidence relevant to the case.

*All doctors are in it for themselves. If yours really did give up all that time for no payment, then all I can say is that there must have been some hidden gain we don't know about.*

(In addition to the less well-hidden fallacy we do know about.)

Aprioristic reasoning is widely used by those whose beliefs have very little to do with reality anyway. The fallacy is the short brush which sweeps untidy facts under a carpet of preconception. It is a necessary household appliance for those determined to keep their mental rooms clean of the dust of the real world. Engraved on the handle, and on the mind of the user, is the legend: 'My mind's made up. Don't confuse me with facts.'

Many of us might be unimpressed with a patent medicine for which the claim was made that recovery proved that it worked, and lack of recovery was proof that more of it were needed. We might point out that the facts were being used to support the medicine, whichever way they turned out. Yet every day precisely the same claim is made for overseas development aid to poorer countries. If there is development, that shows it works. If there is no development, that shows we must give more of it. Heads they win, tails logic loses.

The fallacy of apriorism can also be used to support a preconceived judgement against the evidence. If a politician we support is caught cheating in examinations, or in a compromising position with an intern, these are character-improving situations. They steel him and test him, making him a fitter candidate for office. For anyone else, of course, they would disqualify them from office.

*Since there are no cats in Tibet, this animal here, with the ears of a cat, the tail of a cat, the fur of a cat and the whiskers of a cat, shows that Tibetan dogs are pretty good actors.*

(Not only that, they also catch mice and drink milk from a saucer.)

It is generally unproductive, when using apriorism, to dismiss the facts out of hand as untrue. After all, your audience might have been there to witness them. You will go much further by reinterpreting those facts, showing how they were not what they seemed. Far from contradicting your contention, they really support it.

*I still maintain that the books I recommended were the most popular ones. Of course I don't deny that they were the least read ones in the entire library; but I take that as a sign of their popularity. You see, when a book is really popular, people buy it or borrow it from friends; they don't wait to borrow it from a library.*

(At least the fallacy is popular.)

# *Baculum, argumentum ad*

When reason fails you, appeal to the rod. The *argumentum ad baculum* introduces force as a means of persuasion, and is often resorted to by those who would lose the argument without it.

*It would be better if you told us what we want to know. After all, we wouldn't want your aged mother or your crippled sister to suffer, would we?*

(Probably yes.)

The threatened force does not have to take the form of physical violence. The *argumentum ad baculum* is committed whenever unpleasant consequences are promised for failing to comply with the speaker's wishes. ('If you do not bring us the plans of the new missile, I regret I will be forced to send these photographs to the newspapers.')

The fallacy of the *argumentum ad baculum* lies in its introduction of irrelevant material into the argument. Strictly speaking, it leaves the argument behind, moving on to force as a means of persuasion. While force is undoubtedly effective sometimes in directing courteous attention to the speaker's wishes, its use represents the breakdown and subversion of reason.

The *ad baculum,* alas, performs on the public stage of interna-tional relations. Powerful countries which fail to get their own way by reasoned discussion are not averse to tossing over an *ad baculum* to influence the talks. If even this fails, they toss over something a little larger.

Joseph Stalin was a master of the *ad baculum.* Indeed, he made it his own to such an extent that his name is immortalized in a line of Krushchev which sums up its potency: 'When Stalin says "dance!" a wise man dances.' Stalin himself appears to have taken the view that anyone without force to threaten had no business being involved in international affairs. His famous question, 'How many divisions has the Pope?', was in response to a suggestion that the Pope should take part in an international conference. As Stalin's enemies often discovered, argument is not a very effective counter to an *ad baculum.*

Political parties founded on an idealized view of human nature frequently accuse their rivals of too frequently resorting to *ad baculum* diplomacy. Sir William Browne delivered a well-wrought epigram on the subject:

*The King to Oxford sent a troop of horse,*
*For Tories own no argument but force:*
*With equal skill to Cambridge books he sent,*
*For Whigs admit no force but argument.*

(It would be a close thing today to decide whether it would be harder to find a Tory at Oxford than a literate man at Cambridge.)

You can use the *ad baculum* when you have the force to deploy and can escape the consequences of using it. The law is there to prevent arguments always being won by the stronger, and the many broken bones it would take to determine which was he. But your threats need not be strong physical ones to be effective. Many a speaker has gained his way by threatening to make an intolerable nuisance of himself until his demands were met. The Romans probably destroyed Carthage just to shut up Cato.

# Bifurcation

The presentation of only two alternatives where others exist is called the fallacy of bifurcation. Sometimes known as the 'black and white' fallacy, it presents an 'either/or' situation when in reality there is a range of options.

*If you are not with us, you are against us.*

(Some people might think you partly right. Others might be with you on some things, against you on others. The vast majority probably do not care enough to have an opinion at all.)

Some situations in life have infinite gradations; others offer a straightforward choice. There are many intermediate shades between light and dark, but not all that many things between a boy and a girl. The fallacy of bifurcation consists in taking the limited choice of the second class into situations more properly covered by the first.

*There are two types of people in this world: the rich and the suckers. Do you want to get rich, or are you happy to remain a sucker?*

(In fact there are degrees of richness, as there probably are of sucker-dom. You can be rich by comparison with some, but poor when set alongside others. Suckers, too, seem spread across a continuum.)

The mistake is made by the denial of extra choices. In limiting the field, the perpetrator is leaving out of the discussion material which could well influence the outcome. The fallacy this time is caused not by the intrusion of irrelevant material, but by the exclusion of relevant items.

Bifurcation is used to limit choice. Large political parties employ it to squeeze out smaller ones by denying that they are valid options. Fanatics, for and against, use it to flail the vast mass in between who cannot be bothered. Ideologues use it to classify people into one category or another, rather than admit to the vast range of individual opinions.

One of the more irritating uses of the fallacy of bifurcation occurs in the collection of statistical information. Marketing research polls, along with official forms, can only work by assigning people into broad categories. Information is often requested with the answer 'yes' or 'no' when the individual concerned knows that neither is correct. Personality tests which pose hypothetical situations always grossly underestimate human ingenuity.

Bifurcation often occurs in a dilemma, even though the dilemma itself is a sound form of argument.

*If we import goods we send our jobs abroad; if we export goods we send our property abroad. Since we must either export or import, we lose either our jobs or our property.*

(But it is not a black-and-white choice. We can import some things, export others.)

Lord Nelson uttered the famous cry:

*Westminster Abbey or victory!*

(Overlooking the possibility that he might get both; or the option of St Paul's, where he ended up.)

The greatest use you can make of bifurcation is to offer a choice limited to something very unpleasant or the course you are advocating. Either the audience does what you suggest, or it will be the end of all life on earth as we know it.

*Either we paint the door green, or we will be mocked and ridiculed. People will think we have no taste at all, and we'll become the laughing stock of the whole neighbourhood. I leave the choice up to you; I'm not trying to influence your decision one way or the other.*

You must learn to introduce what you consider to be the only possible choice by saying: 'Well, ladies and gentlemen, it seems we have two possible choices . . .'

# Blinding with science

Science enjoys an enormous prestige because it has got so many things right. In the popular imagination, the dedicated scientist in his white coat is a fount of real knowledge as opposed to mere opinion. The fact that he is using that knowledge to make Frankenstein monsters scarcely diminishes the respect for his pronouncements. Many people, anxious to invest their own views with the authority of the scientist, don the white coat of scientific jargon in an attempt to pass off their own assertions as something they are not.

The fallacy of blinding with science specializes in the use of technical jargon to deceive the audience into supposing that utterances of a scientific nature are being made, and that objective experimental evidence supports them.

*The amotivational syndrome is sustained by peer group pressure except where achievement orientation forms a dominant aspect of the educational and social milieu.*

(Which means roughly that people don't work if their friends don't, unless they want to get on. Now this may be true or false, but many are daunted from challenging what is dressed up to look like an expert view.)

The white coat of technical jargon is so dazzlingly clean (never having been tainted by any real scientific work) that it blinds the audience to the true merits of what is being said. Instead of evaluating contentions on the basis of the evidence marshalled for and against them, the audience recoils from the brilliance of the jargon. The fallacy is committed because this irrelevant material has no place in the argument. Just as loaded words try to prejudice a case emotionally, so does pseudo-scientific jargon try to induce an unearned respect for what is said. The proposition is the same, whatever the language; and use of language to make its acceptance easier is fallacious.

Although blinding with science can be used in any argument, many will recognize the special domain of this fallacy as the subjects which like to consider themselves as sciences, but are not. Science deals with things from atoms to stars at a level where individual differences do not matter. The scientist talks of 'all' rolling bodies or whatever, and formulates general laws to test by experiment. The trouble with human beings is that, unlike rolling bodies, the individual differences do matter. Often, again unlike rolling bodies, they want to do different things. Although this might prevent us from being scientific about human beings, it does not stop us pretending to be so. What we do here is to add the word 'science' onto the study, giving us 'economic science', 'political science' and 'social science'. Then we dress them in that white coat of scientific language, and hope that no one will notice the difference.

*The transportational flow charts for the period following the post-meridian peak reveal a pattern of decantation of concentrated passenger units in cluster formations surrounding the central area.*

(You could spend years formulating laws to predict this, and might even be in the running for a Nobel prize. Just remember never to mention that people are coming into town to have a bite to eat, followed by a movie or a show . . .)

The first rule for using this fallacy is to remember to use long words. ('When the pie was opened, the birds *commenced* to sing.') Never use a four-letter word, especially if you can think of a 24-letter word to take its place. The jargon itself is harder to master, but a subscription to *New Society* is a good investment. Remember that the basic function of words is to prevent communication. Their real task is to transform what is banal, trivial and easily refuted into something profound, impressive and hard to deny.

> *The small, domesticated carnivorous quadruped positioned itself in sedentary mode in superior relationship to the coarse-textured rush-woven horizontal surface fabric.*

(With its saucer of milk beside it.)

The fallacy of blinding with science is well worth the time and trouble required to master it. The years of work at it will repay you not only with a doctorate in the social sciences, but with the ability to deceive an audience utterly into believing that you know what you are talking about.

# The bogus dilemma

Quite apart from the casual use of the term to describe a difficult choice, the dilemma is also the name of an intricate argument. In a dilemma, we are told the consequences of alternative actions, and told that since we must take one of the actions, we must accept one of the consequences. A Greek mother told her son who was contemplating a career in politics:

> *Don't do it. If you tell the truth men will hate you, and if you tell lies the gods will hate you. Since you must either tell the truth or tell lies, you must be hated either by men or by the gods.*

The dilemma is a valid form of argument. If the consequences described are true, and if there really is a straight choice between them, then one or other of the consequences must follow. Very often, however, the information given is incorrect, and the choice is not as limited as is made out. In these cases the dilemma is bogus. The bogus dilemma is the fallacy of falsely or mistakenly presenting a dilemma where none exists.

In the above example, the son has several possible replies. He can claim that the dilemma is bogus by denying that the consequences follow – this is called 'grasping the dilemma by the horns'. He can simply deny that men will hate him if he tells the truth: on the contrary, he might claim, they would respect him for it. The alternative statements about consequences are called the 'conjuncts', and it is enough to show that one or both is false to label the dilemma as bogus. As another option, he might show that the choice is false. This is called 'going between the horns', and consists of showing that other choices are possible. Instead of limiting himself to truth or lies, he might be truthful at some times, deceitful at others. He might make statements which contain elements of both truth and falsehood. The dilemma is shown to be bogus if the choice, which is called the 'disjunct', is not an exhaustive one. A third way of dealing with a dilemma is to rebut it. This is an elegant technique which requires an equally ferocious beast to be fabricated out of the same elements as the original one, but sent charging in the opposite direction to meet it head-on. In the above example, the youth replied:

*I shall do it, mother. For if I tell lies, men will love me for it; and if I tell truth the gods will love me. Since I must tell truth or lies, I shall be beloved of men or gods.*

(This is so pretty that when one sees it done in debate, there is an urge to throw money into the ring.)

Protagoras, who taught law among other things, dealt with a poor student by agreeing to waive the fee until the man had won his first case. As time went by, and there was no sign of the youth taking on a case, Protagoras sued him. The prosecution was simple:

> *If the court decides for me, it says he must pay. If it decides for him, he wins his first case and must therefore pay me. Since it must decide for me or for him, I must receive my money.*

The youth had been a good student, however, and presented the following defence:

> *On the contrary. If the court decides for me, it says I need not pay. If it decides for Protagoras, then I still have not won my first case, and need not pay. Since it must decide for me or for him, either way I need not pay.*

(The judge had a nervous breakdown and adjourned the case indefinitely. He thereby proved the disjuncts false, and escaped between the horns of both dilemmas.)

The fallacy in the bogus dilemma consists of presenting false consequences or a false choice, and it will be of most use to you in situations where decisions which you oppose are being contemplated. Quickly you step in, pointing out that one of two things will happen, and that bad results will follow either way:

> *If we allow this hostel for problem teenagers to be set up in our area, either it will be empty or it will be full. If it is empty it will be a useless waste of money; and if it is full it will bring in more trouble-makers than the area can cope with. Reluctantly, therefore . . .*

(Cross your fingers and hope there are no students of Protagoras on the committee.)

# Buzzwords

A buzzword is not in itself a fallacy. It is simply a word or phrase used to impress, or one that is fashionable. The fallacy enters when the use of the word makes the listener or reader give more weight to what is being communicated than the arguments or facts themselves sustain. The person hearing the claim supported by a buzzword might suppose it more likely to be true than it is in fact, or that it merits more importance than it would command without the use of the buzzword.

Buzzwords come and go. Everyone (except perhaps readers of the UK's Daily Telegraph) likes to be up to date or to be seen to be up to date, and use of a current buzzword can achieve this. This has a great deal to do with being fashionable, and very little to do with being right. A judicious buzzword can command attention and persuade the listener that the user is on top of current thinking and developments. It can even persuade people that the user knows what they are talking about.

A buzzword has to be current or it no longer merits the name. One that is no longer in fashion slips from being a buzzword and joins the collection of yesterday's jargon – jargon that makes users look hopelessly outmoded. There is thus an element of the *argumentum ad novitam* about buzzwords; you are expected to esteem them because they are new. Buzzwords are a special case of the more general fallacy.

*My plan will enhance the public's perception of our corporate social responsibility.*

(It will make the firm look caring, thus enabling it to sell more products.)

Even 'corporate social responsibility' is looking decidedly moth-eaten, and almost ready to settle into a retirement home alongside 'empowerment' and 'stakeholders'. Brash new guys have moved in to take over their territory.

> *I propose we incorporate big data into our company's forward planning.*

> ('Big data' is hard to resist as a buzzword because not many people would favour using less information. Whatever it is that is being proposed, 'big data' carries it further forward than its merits would.)

To use this fallacy you obviously need to be on top of currently fashionable (but fairly empty) words and phrases. Do bear in mind that buzzwords have a limited shelf life. They pepper presentations as people try to impress, but like yesterday's cakes they soon lose their freshness and their appeal. Even the buzzwords in this book might well have passed their sell-by date by the time it is published.

You need a repertoire of the new ones. Unless you attend the board meetings of a major corporation this can be difficult. You should frequent bars used by City traders and stand close enough to pick up some of the current buzzwords, but do this early in the evenings when the nonsense phrases are at least coherent. Failing this, you can learn what buzzwords will augment your argument by scanning the bookracks at airports to see what titles are persuading people to buy another business book that will tell them no more than the others did.


# *Circulus in probando*

*Circulus in probando* is a specialized and very attractive form of the *petitio principii*. It consists of using as evidence a fact which is authenticated by the very conclusion it supports. It is thus arguing in a circle.

*'I didn't do it, sir. Smith minor will vouch for my honesty.'*
*'Why should I trust Smith minor?'*
*'Oh, I can guarantee his honesty, sir.'*

(Any teacher who falls for that one deserves to be suspended by his thumbs from two hypotheticals.)

The *circulus* is fallacious for the same reason as is its larger cousin, *petitio.* It fails to relate the unknown or unaccepted to the known or accepted. All it gives us is two unknowns so busy chasing each other's tails that neither has time to attach itself to reality.

*We know about God from the Bible; and we know we can trust the Bible because it is the inspired word of God.*

(A circle in a spiral, a wheel within a wheel.)

As with the *petitio,* its close relative, the *circulus* is often found building a cosy little nest in religious or political arguments. If there really were convincing proofs of particular religions or ideologies, it would be much more difficult for intelligent people to disagree about them. In place of cast-iron demonstrations, *petitio* and *circulus* are often called upon to serve.

The same could even be said of science. How do we know that our so-called scientific knowledge is no more than one giant *circulus*? When we perform scientific experiments, we are assuming that the rest of our knowledge is good. All we are really testing is whether the new theory under examination is consistent with the rest of our theories. At no point can we test any of them against some known objective truth. After all, even the theories about what our senses tell us are in the same predicament. It all comes down to saying that science gives us a consistent and useful look at the universe through the ring of a giant *circulus.*

You will find it difficult, however, to use the prestige of science in support of your own use of the *circulus.* He is too easily spotted for effective application in argument, being rather less wily than his big cousin, *petitio.*

> '*I have the diamond, so I shall be leader.*'
> '*Why should you get to keep the diamond?*'
> '*Because I'm the leader, stupid.*'

The more likely your conclusion is to be acceptable for other reasons, the more likely are you to get away with a *circulus* in support of it. When people are already half-disposed to believe something, they do not examine the supporting arguments as closely. That said, *circulus* should be reserved for verbal arguments where memories are short.

> '*I'm asking you to do this because I respect you.*'
> '*How do I know that you respect me?*'
> '*Would I ask you to do it otherwise?*'

(If you want to do it, you'll believe it.)

The intelligent reader might suppose that fallacies such as *circulus* are too obvious to be more than debating tricks. Surely they could never seriously distort decisions of state by slipping through the massed ranks of civil servants, government committees and the cabinet? Not so. A major policy of Britain's government in the 1960s, adopted after the most serious public debate, was based upon a relatively obvious *circulus in probando.* This was the National Plan, an exercise in (then fashionable) national economic planning. Firms were asked to assume a national growth-rate of 3.8 per cent, and to estimate on that basis what their own plans for expansion would be. These various estimates were added up by the government, which concluded that the combined plans of British industry suggested a growth-rate of 3.8 per cent!

The National Plan was valueless then and subsequently, except to connoisseurs of logical absurdity lucky enough to snap up remaindered copies of it in secondhand bookshops.

# Collective guilt

Guilt is something that pertains to individuals, not to institutions or categories. There may be many guilty individuals, as when a gang of people participates in rape or looting. The leaders of a country can be guilty of crimes or misdeeds, but not the country itself. The same applies to a nationality or a race or an ethnic group. Individuals in the group might be guilty, but it is a fallacy to suppose that the guilt pertains collectively to the group.

*America should apologize to Africa for the slavery it practised, and pay reparations.*

No-one living in present-day America took part in that slavery. Some citizens have ancestors who did, and probably many more have no such ancestors. In no sense is America collectively guilty of slavery (though you could make a case that a four-year civil war and 600,000 dead represented a pretty handsome apology).

*Britain was guilty of enriching itself through colonialism and ought to compensate its former colonies accordingly.*

Many people think that Britain became rich by creating wealth, not by taking it from others. But even if some did, few if any are alive now. To suggest that present British citizens, including those who went there later and made Britain their home, have an obligation of guilt is to commit the fallacy. Crimes and misdeeds are committed by people, not by nationalities.

A headmaster once had to preside over an enquiry into an incident in which one pupil had knifed another. As the meeting began with the parents present, the educational psychologist began, 'Headmaster, we should not see this as the boys' problem. It is society's problem and in some sense we must all bear part of the guilt.' At this point the headmaster gathered up his books declaring, 'Well I'm not bloody guilty!' as he left the room. He never allowed an educational psychologist in his school again. His reply was the perfect response. When charged with collective guilt, deny it individually.

The easiest way to use the fallacy yourself is to accuse an opponent of being guilty of something because he or she belongs to a class that is or was morally guilty, even though your opponent is personally innocent. This is especially good in the UK because no-one can resist generalizations about class.

*'It is you middle class people who are guilty of spreading philistinism by preferring chocolate box covers to truly dynamic modern art.'*

(Even if the guy founded the Guggenheim Museum, he is still tarred with the brush of his class's collective guilt.)

# The complex question (*plurium interrogationum*)

*Plurium interrogationum,* which translates as 'of many questions', is otherwise known as the fallacy of the complex question. When several questions are combined into one, in such a way that a yes-or-no answer is required, the person they are asked of has no chance to give separate replies to each, and the fallacy of the complex question is committed.

*Have you stopped beating your wife?*

(If 'yes', you admit you were. If 'no', then you still are.)

This might seem like an old joke, but there are modern versions:

*Did the pollution you caused increase or decrease your profits?*

*Did your misleading claims result in you getting promoted?*

*Is your stupidity inborn?*

All of them contain an assumption that the concealed question has already been answered affirmatively. It is this unjustified presumption which constitutes the fallacy. Many questions may be asked, but if the answer to some is assumed before it is given, a *plurium interrogationum* has been committed.

A common version of the fallacy asks questions beginning 'who' or 'why' about facts which have not been established. Even oldies such as 'Who was the lady I saw you with last night?' and 'Why did the chicken cross the road?' are, strictly speaking, examples of this fallacy. They preclude answers such as 'There was none', or 'It didn't.'

> *Why did you make your wife alter her will in your favour? And why did you then go along to the chemist to buy rat poison? Why did you then put it into her cocoa, and how did you do it without attracting her attention?*

(Attempt not more than three questions.)

The inhabitants of the world of the *plurium* are a puzzled lot. They can never understand why we tolerate television reporters who echo anti-patriotic propaganda, how we can curb drug abuse in our schools, or why it is that so many unemployable people are produced by our universities and colleges. The advertisers of that world want to know whether our families are worth the extra care that their product brings and if we are glad we chose their brand of shampoo.

In the real world none of these questions would be regarded as valid until the facts they depend on had been established. The complex question has to be broken into simpler ones; and often the denial of the fact presumed invalidates the larger question altogether.

A variety of complicated genetic or evolutionary explanations could be advanced to explain why the adult human female has four more teeth than the adult male. None of them would be as effective as counting along a few jaws and denying the fact.

*Plurium interrogationum* is very effective as a means of introducing the semblance of democracy into the domestic scene. It enables you to give children a choice over their destinies:

*Would you prefer to go to bed now, or after you've finished your cocoa?*

*Do you want to put your bricks in the box, or on the shelf?*

(Beware, though. After about ten years this will come back to you as: *Mum, would you prefer to buy me a disco deck or a motorbike for my birthday?*
   He who sows the wind . . .)

# Composition

The fallacy of composition occurs when it is claimed that what is true for individual members of a class is also true for the class considered as a unit. Some nouns can be taken to refer either to the thing as a whole or to the various parts which make it up. It is fallacious to suppose that what is true of the parts must also be true of the new entity which they collectively make up.

*This must be a good orchestra because each of its members is a talented musician.*

(Each individual might be excellent but totally unable to play in unison with colleagues. All of these virtuosos might be far too busy trying to excel personally to play as an effective team.)

Many a football manager has similarly transferred in many first-class players, only to find himself transferred out. Unless they can work as a team, it is easier to get the manager out of the ground than the ball into the net.

*I have gathered into one regiment all of the strongest men in the army. This will be my strongest regiment.*

(I doubt it. The strength of a regiment depends on such factors as its morale and its teamwork, not to mention its speed, its ability to operate with minimal supplies, and similar attributes.)

The fallacy arises from a failure to recognize that the group is a distinct entity of which things can be said which do not apply to individual persons. Evidence advanced to attest to the qualities of the members is therefore irrelevant to an appraisal of the group.

Americans are particularly vulnerable to this fallacy because their grammar makes no distinction between the collective entity and the individuals within it. It seems to be universal in the American language to use singular verbs for collective nouns, regardless of whether the members or the group are being referred to.

In English we would say 'the crew is a good one', referring to it as a separate entity, but 'the crew are tired', if we are speaking of its members. In American one uses the singular verb in both cases, losing an important distinction.

*If everyone in society looks after themselves, then our society will be one that looks after itself.*

(It will certainly be a society of people who look after themselves; but maybe society has aspects which need to be looked after by people acting in concert.)

A variant of the fallacy of composition covers cases in which things which are true for individuals become untrue if they are extrapolated to cover the whole group.

*Farmers benefit from price supports on beef; shoemakers gain from price supports on shoes, and so on. Obviously the whole economy would benefit if all products were subsidized.*

(The point is that farmers and shoemakers only benefit if they are in a small group which benefits at the expense of everyone else. If the principle is extrapolated, everyone receives the subsidies, everyone pays the taxes to fund them, and everyone loses out to the bureaucrats who administer the transfers.)

Society, indeed, provides the best place to use the fallacy with intent to deceive. You should attribute all kinds of sympathetic qualities to the people in our country. An audience of your countrymen will have no difficulty in attesting to the truth of them. When you slide in a surreptitious fallacy of composition to urge the same for society as a unit, they will be reluctant to let go of the good qualities they just claimed.

*We all know that the average Briton is noted for a warmhearted generosity. That is why our society has to increase the rights of the old, the sick, the unemployed, and those in less developed countries.*

(These actions might be worthwhile, but are only generous when done by *individuals*. To take money away from people to give to others actually diminishes their opportunity to be generous.)

You might just as well try: 'Irishmen tend to die young, you know. I'm surprised the country is still going.'

# Concealed quantification

When statements are made about a class, sometimes they are about all of the members of it, sometimes about some of them, and at other times it is not clear which is referred to. The fallacy of concealed quantification occurs when ambiguity of expression permits a misunderstanding of the quantity which is spoken of.

*Garage mechanics are crooks.*

(What, all of them? It does not say, but there is a big difference. If it refers to all of them, then to talk to one is to talk to a crook. Although many motorists may have their convictions, few of the garage mechanics do.)

Very often the quantification is concealed because it sounds rather lame to make bold statements about *some* of a class. 'All' is much better, but probably untrue. Rather than be limited by such a technicality, a speaker will often leave out the quantity in the hope that 'all' will be understood. Someone might commiserate with a distraught parent by telling them: 'Teenagers are troublesome.' This can be accepted as 'some are,' or even 'many tend to be so', but it could also be taken to mean that one has only to find a teenager to locate a troublesome person. This may not have been intended, however plausible it sounds. The fallacy comes with the ambiguity. The statement can be accepted with one meaning, yet intended with another. Of course, very different conclusions can be drawn from the two meanings.

*It is well known that members of the Campaign for Nuclear Disarmament are communists.*

(It is indeed, but not all of them, as seems to be implied. Even if some are communists, there is still room for others motivated by sincerity or stupidity.)

The fallacy is widely used to condemn whole groups on the basis of some of their members.

*Subversives teach at the Open University.*

(This could mean that some do, but is unlikely to mean that all subversives are so employed. It could even be taken as telling us that *only* subversives teach there. The quality of the average BA would vary enormously, depending on which were true, as indeed might the course content.)

Concealed quantification can also be a prelude to tarring an individual with the characteristics of the group to which he belongs by hiding the fact that they apply only to some of that group.

*Have you ever noticed that bishops are fat? I suppose now that Johnson has been raised to a bishopric he'll expand a bit himself.*

(Weight and see.)

You should use concealed quantification to make a weak case look stronger than it is. If you are trying to sow doubts about a person, you can use their membership of some group to cast general aspersions about them. Make reasonable-sounding statements which are true of some, and allow your audience to supply the 'all' or the 'only' which are needed to brand him as well.

*I don't think we should hire Thomson. I see he's a keen fisherman. Idlers take up fishing, so I think it's a very bad sign.*

(The audience take the bait, make it *'only* idlers', and Thomson is already hooked.)

# Conclusion which denies premises

The conclusion which denies its premises is one of the 'oh-dear-I-forgot-what-I-started-to-say' fallacies. It starts by maintaining that certain things must be true, and ends up with a conclusion which flatly contradicts them. If the conclusion is not consistent with the arguments used to reach it, then somewhere there is a hole in the reasoning through which the logic has slipped silently away.

*'Son, because nothing is certain in this world we have to hold on to what experience tells us.'*

*'Are you sure, Dad?'*
*'Yes, son. I'm certain.'*

The fallacy is identified by the inconsistency. If the conclusion contradicts the premises, at least one of them must be wrong. This means that our conclusion is either false itself, or derived from false information.

The conclusion which denies its premises constantly slips uninvited into religious arguments. People are so used to thinking of divine beings as exceptions to every rule that they tend to use the word 'everything' when they mean 'everything except God.'

*Everything must have a cause. That, in turn, must result from a previous cause. Since it cannot go back for ever, we know that there must be an uncaused causer to start the process.*

(But if everything must have a cause, how can there be such a thing as an uncaused causer?)

The fallacy has a most distinguished history, being used (although not identified as such) by Aristotle and Thomas Aquinas among many others. It has many faces. The 'uncaused causer' can be a 'first cause,' or even a 'first mover'. It can be reworded in many ways, but never without fallacy.

Attempts to make a divine being the allowable exception to the original claim usually beg the question or subvert the argument, 'Everything in the universe must have a cause outside of itself . . .' The intention is clearly to establish a cause which is outside of the universe and therefore needs no cause to account for it. Unfortunately, the rewording admits several faults.

1    The new version is more complex and is not obviously true.

2    The universe is not *in* the universe, it *is* the universe.

3    'Everything in the universe' *is* the universe.

This allows us to translate the opening line as: 'The universe must have a cause outside of itself.' Given such an assumption, it is hardly surprising that we go on to prove it.

There are many simpler versions in popular currency, none of them free from the basic inconsistency of allowing the preferred answer to be the one permitted exception.

> *No matter how many stages you take it back, everything must have had a beginning somewhere. God started it all.*

(He, presumably, did not have a beginning somewhere.)

> *Nothing can go on forever. There must have been a god to start it.*

(One who goes on forever, of course.)

When using the conclusion which denies premises, you should bear in mind three things. First, the more distance there is between your opening line and your conclusion, the less likely are your audience to spot the contradiction. Second, they will often allow a speaker to make statements about 'everyone' without applying them to the speaker himself. Third, if your conclusion is about things which are usually admitted to have exceptional properties, your fallacy has a better chance of escaping detection.

> *Never believe what people tell you about patent medicine; they are always liars. It is because you know that I am truthful about this that you will know I am also telling the truth when I tell you that my snake-oil here is the most remarkable . . .*

(Remarkable.)

# Contradictory premises

No matter how good the logic may be, you cannot rely on an argument which has certain falsity built into it. For a sound argument true premises are needed, as well as valid logic. The problem with contradictory premises is that they cannot both be true. If one is true, the other must be false, and vice versa. In other words, we can be certain that at least one of them must be false, and cannot therefore generate a sound argument.

> *Everything is mortal, and God is not mortal, so God is not everything.*

(This might look like an argument against pantheism, but it is in fact an argument against common sense. Since the premises contradict each other, one must be false. This makes any conclusion unreliable.)

The fallacy is an interesting one because it permits the logic to be valid. It usually amazes non-logicians to hear that with inconsistent premises any conclusion, no matter how irrelevant, can be validly inferred. Logicians, however, do not use the word 'valid' to mean 'sound', If there is known falsity built in, as there must be with contradictory premises, then it matters not how good the logic is: the argument is not sound.

This is the fallacy which enables us to prove that the moon is made of green cheese. The proof is quite complicated, but quite fun:

> We are given two premises, that milk is white, and milk is not white. If 'milk is white' is true, then it is also the case that 'Either milk is white or the moon is made of green cheese' is true. (This is correct.) Since we are also given that milk is not white, the second alternative must be true, namely that 'the moon is made of green cheese'.

There is nothing wrong with the logic. The known falsity in the contradictory premises can be used to establish anything, including a rather smelly satellite.

It is difficult to use the fallacy of contradictory premises in every-day argument, because your audience will normally see that you are contradicting yourself. What you can do, however, is to use contradictions which are normally accepted in loose speech, and proceed to wrap them up in tight logic.

*He's a real professional, but a bit of an amateur at times.*

(It sounds acceptable enough, but remember that from it you can literally prove that the moon is made of green cheese.)

# Crumenam, argumentum ad

The *argumentum ad crumenam* assumes that money is a measure of rightness, and that those with money are more likely to be correct. 'If you're so right, why ain't you rich?' is the common form, but it translates poetically as the belief that 'truth is booty'.

There have been branches of Christianity which held that worldly success could be taken as a mark of divine favour; and there have been constitutions which loaded the franchise to the advantage of those with wealth and property.

*I note that those earning in excess of £100,000 per year tend to agree with me.*

(Maybe so. He might have added that right-handed people disagreed with him, that 6-foot-tall people agreed, and that those with hazel eyes were evenly divided. These have about as much to do with being right as money does.)

The fallacy in the *argumentum ad crumenam* is, of course, that wealth has nothing to do with it. It is a sweet and fitting thing to make a lot of money. It is also a sweet and fitting thing to be right; but only an undistributed middle can relate the two because of this.

Behind the *argumentum ad crumenam* there lies the vague feeling that God would not allow people who were wicked and wrong to scoop the pool of life's goodies. We know that money isn't everything, but we suspect, deep down, that it is 90 things out of 100, that it will buy nine of the remaining ten, and even make the absence of the remaining one tolerably comfortable.

*Surely a man who can make £60 million in a year by recording four songs cannot be all wrong?*

(He can.)

*The world's most expensive beer . . .*

(But it makes you no more drunk than does the cheapest.)

There are limited and artificial situations in which money is the measure of right.

*The customer is always right.*

(This is because the customer has the money. It is true in America; but in Britain the convenience of the shopkeeper often comes first, and in France or Germany it always does.)

In the field of tipping, money can often bring right in its wake.

*'Cabbie, get me to the airport by ten o'clock!'*
*'This cab ain't got wings, mister.'*
*'Here's £20 if you make it.'*
*'Stand by for take-off!'*

*'My friend wants to know where Big M was last night.'*
*'Who's your friend?'*
*'He sent his picture.'* [waves banknote]
*'You can tell Sir Edward Elgar that Big M was at Molly's bar.'*

A version of the *argumentum ad crumenam* helped in the suc-cess of the Industrial Revolution. The belief that the virtues of thrift, perseverance and hard work are rewarded by wealth led naturally to its converse, that worldly goods were the hallmark of virtue. A society in which one needs to make money to be respected for moral worth is probably conducive to an expanding economy.

Your own use of the fallacy is best reserved for situations where you personally can ensure that money not only talks, but positively monopolizes the conversation.

> *'I say we do it this way, and I own 60 per cent of the shares in this company.'*

> [chorus] *'You're right, J.G.!'*

This differs only in degree from the junior version:

> *'I say it was a goal, and it's my football.'*

## *Cum hoc ergo propter hoc*

The *cum hoc* fallacy assumes that events which occur together are causally connected, and leaves no room either for coincidence, or for the operation of an outside factor which separately influences those events.

> *A tourist met a Spanish peasant and his wife on a train. They had never seen bananas before, so he offered one to each of them. As the farmer bit into his, the train entered a tunnel. 'Don't eat it, Carmen,' he shouted, 'They make you blind.'*

Like the *post hoc* fallacy which links events because they occur consecutively, the *cum hoc* fallacy links them because they occur simultaneously. It is a fallacy because of its unwarranted assumption that either of the events would not occur without the other one.

Things are happening all the time. Scarcely a day goes by without rain, electricity bills, televised show-jumping, or the *Guardian* newspaper. It is inviting to link these endemic discomforts with simultaneous events, and conclude that they are all somehow connected. In primitive societies such assumptions are made routinely, and one of the jobs of the witchdoctor is to sort out which actions are linked with various consequences. In our society, alas, life is more complicated.

The field of statistics provides a natural habitat for the *cum hoc* fallacy to lurk undetected. Indeed, a whole branch of statistics called regression analysis is devoted to measuring the frequency and extent of simultaneous events, and to calculating the probability of their being linked. Correlation coefficients are produced, with percentages attached showing the likelihood that mere chance was involved. Statisticians routinely offer us relationships with a 95 per cent or a 99 per cent probability that 'more than chance is involved'.

*A statistician looking over figures for pupil performance was astounded to discover in the 7–12 age-group that neatness of handwriting matched with size of shoe. He checked the figures for hundreds of children, but it was quite clear. Neat handwriting correlated with large feet, with 99 per cent probability that this was not mere chance.*

(A teacher later told him that this was because older children tended to write more neatly. Being older, they tended to have bigger feet.)

Most disciplines which involve human measurement, including economics and sociology, find *cum hocs* scattered liberally on their domain. The reason for this is that we do not really know what makes human beings act, so we look at their actions after the fact and try to relate them to other events. The *cum hoc* tares grow up with the wheat of genuine insights.

*Elections make people spend. The figures are clear. Spending always goes up in an election year.*

(Could it be that governments seeking re-election tend to keep taxes down in election years, and that people, in consequence, have more to spend?)

Deliberate use of the *cum hoc ergo propter hoc* is best made with the support of reams of statistical information. Your audience, bemused by the figures, rarely have any of their own to set against you. They can be made even more disposed to accept the link which you are proposing if you cite the authority of leading figures in the social sciences. This is easy. There is nothing so absurd that it has not been attested to by such people. It helps to be selective in your use of information.

*Gun ownership is a major cause of violent crime. The prevalence of guns in the US matches the high rates for crimes of violence. When violence is contemplated, the guns are all too available.*

(Excellent; but remember not to mention Switzerland, where almost every household has a gun as part of military training. Switzerland has low rates for violent crime, and the guns are almost never used.)

A US legislator recently noted that a high crime-rate correlated with a high prison population, and suggested that the prisoners be released in order to cut the crime figures.

For use of the fallacy in print, simply juxtapose articles. Study the front pages to see how it is done.

MARK TWAIN COMES TO TOWN
ASCOT GOLD CUP STOLEN

# Damning the alternatives

In cases where there is a fixed and known set of alternatives, it is legitimate to establish the superiority of one by showing all of the others to be inferior. However, in cases where the alternatives are not fixed or known, and where absolutes rather than comparatives are sought, it is a fallacy to suppose that we argue for one by denigrating the alternatives. The fallacy is that of damning the alternatives.

*Hawkins' theory has to be the right answer. All the others have been proved hopelessly wrong.*

(And his may be proved wrong tomorrow.)

Even where there are only two alternatives, we cannot show that one is good by showing that the other one is not. Both might be equally bad. The same applies for larger groups.

*Chelsea is a really great team. Look at Liverpool and Manchester United; they are both useless.*

(Other teams not taken account of might enter the reckoning. Even so, if Liverpool and Manchester United were bad, it would not prove Chelsea good. It might be that all football teams are absolutely terrible.)

The fallacy occurs because in leaving out the performance of alternatives not referred to, we exclude material which might be relevant to a decision. Second, by introducing material which denigrates others in cases where a simple judgement is required, we bring in irrelevant matter.

Damning the alternatives is the fallacy of the partisan. Anxious to elevate his own village, nation, team, church, occupation, race or class, he thinks he does so by running down the others. Rupert Brooke used the fallacy for humorous effect in his famous poem, 'The Old Vicarage, Grantchester'.* Amongst the praise for Grantchester itself are sandwiched adverse comments on the other villages in the area. He tells us:

> For Cambridge people rarely smile,
> Being urban, squat and packed with guile . . .
> Strong men have run for miles and miles
> When one from Cherry Hinton smiles . . .
> Strong men have blanched and shot their wives
> Rather than send them to St Ives.

In British elections it is considered bad form for a candidate to promote his own cause by castigating his opponents; he lets his election agent do it instead. In the USA there is no such compunction:

> *You takes your choice: a convicted rapist, an adulterer, a practising pervert, an embezzler and me.*

(The candidates tend to be more exotic in the USA; this might explain it.)

The fallacy will give you hours of innocent fun (and a fair amount of guilty fun) in running down the alternatives to what you are proposing. We appear to have a kind of double vision which leaves us short-sighted on virtue but hawk-eyed for faults. To you this is but opportunity. When you pick on a couple of alternatives and expose their imperfections, the audience will be turning those defective eyes away from your own proposal. They will assume that you would not run down everything

---

*Rupert Brooke, 'The Old Vicarage, Grantchester', in Brooke, *Collected Poems* (London: Sidgwick & Jackson, 1918).

else as mean, foolish, wrong and wicked if your own ideas were no better. They will be mistaken.

*No design for a new building ever meets with universal approval, but look at the alternatives: a glass-fronted matchbox, something with all the pipes on the outside, or a moulded concrete monstrosity.*

(Whereas the one you approve of lets in water, sheds tiles on passers-by and needs a king's ransom to maintain. But they won't see that if you keep them focused on the damned alternatives.)

# Definitional retreat

A definitional retreat takes place when someone changes the meaning of the words in order to deal with an objection raised against the original wording. By changing the meaning, he turns it into a different statement.

*'He's never once been abroad.'*
*'As a matter of fact, he has been to Boulogne.'*
*'You cannot call visiting Boulogne going abroad!'*

(What can you call it then? How about calling it 'sitting on a deck-chair at Blackpool?')

Words are used with conventional meanings. If we are allowed to deal with objections to what we say by claiming that they mean something totally unusual, rational discourse breaks down altogether.

The fallacy in a definitional retreat lies in its surreptitious substitution of one concept for another, under the guise of explaining what the words really mean. The support advanced for the one position might not apply to its substitute. (*'When I said I hadn't been drinking, officer,*

*I meant that I hadn't had more than I get through in a normal social evening.'*)

The definitional retreat allows someone beaten in an argument to save face by claiming that he was really putting forward a totally different view. It also allows for a possible exception to be eliminated by a more restrictive interpretation.

> *'You have no experience of dealing with terrorism.'*
> *'Well, I did act as anti-terrorist adviser to the governments of Malaysia and Singapore, and I spent four years at the US anti-terrorist academy.'*
> *'I meant you have no experience of dealing with terrorists in* England.'

(He should have made it *Scunthorpe,* to be even safer.)

> *'When I said that we were ruled by tyrants, I was naturally refer-ring to the tax-collectors and administrators, rather than to Your Majesty.'*

Definitional retreat is a favourite recourse of philosophers. Their proposed definitions of 'virtue', 'the good', and even of 'meaning' itself, are set up like wickets for their colleagues to bowl at. When the occasional googly scatters the stumps, instead of walking back gracefully to the pavilion, the philosopher is more likely to re-erect the stumps in a slightly different place and show that the ball would not have hit them in that position.

A passage from Lewis Carroll sums it up:

> 'There's glory for you!'
>     'I don't know what you mean by "glory," Alice said.
>     Humpty Dumpty smiled contemptuously. 'Of course you don't – till I tell you. I meant "there's a nice knock-down argument for you!"'
>     'But "glory" doesn't mean "a nice knock-down argument",' Alice objected.

'When *I* use a word,' Humpty Dumpty said in rather a scornful tone, 'it means just what I choose it to mean – neither more nor less.'*

The UK's finance ministers are no less skilled. They have vast numbers of Treasury officials whose sole purpose is to redefine words like 'growth', 'investment', 'spending' and 'business cycle'.

When you marshal your own arguments into a timely definitional retreat, it is advisable to claim a meaning for the words which is at least plausible. It should have some authority behind the usage. One good way is to slip into a technical vocabulary when you started out using ordinary speech.

*Of course, I was using 'expectation' as statisticians do, multiplying the probability of the returns by their size. I didn't mean it in the sense that we* expected *anything to happen.*

(Except, perhaps, for a fish wriggling artfully off a hook.)

A useful device to provide covering fire for a definitional retreat is the presumption that everyone understood your second meaning all along, and only your critic has been so finicky as to ignore it:

*Everybody knows that when we talk of trains being punctual, we use the railway definition of being within ten minutes of the timetable.*

(They do now, anyway.)

# In denial

It is not a fallacy to report that someone denies something. They say they didn't do it, and they may or may not be speaking the truth.

---

*Lewis Carroll, *Through the Looking Glass* (London: Macmillan, 1927), pp. 124–5.

But to say they are 'in denial' means something different. It conveys the message that although we know that they did it, they nonetheless refuse to admit it. How do we know they did it? Supporting material to establish that certainty is nowhere contained within the assertion; guilt is simply assumed. We can assume that the party said to be 'in denial' has not confessed. Denial goes with many things, but confession is not one of them.

There is a limited case where 'in denial' might be used, and that is where someone has faced trial and been convicted by due process. If a jury has heard the evidence for and against and has reached a guilty verdict, we may say that the person has been proven guilty, and that if they refuse to admit their guilty, they are 'in denial.' Do note that even here there is no certainty. If the guilty verdict was reached in a country where a fair trial is a rarity (and I couldn't possibly give you an example of one), the party convicted after a travesty of a trial can still plead their innocence without necessarily being 'in denial.' And even in a country which administers reasonably fair justice, the person might just have been wrongly convicted.

The problem is that 'in denial' implies certain knowledge of a guilt that in most cases has not been established beyond any reasonable doubt.

*'He's in complete denial about his relationship with his secretary.'*

(This implies that even though we all know he's having a bit on the side, he refuses to own up to it. It might be fairer to say that although many people suspect him of an affair, he denies it, but we want to convict him without a fair hearing, and 'in denial' does that.)

Sometimes 'in denial' is used to mean that someone refuses to recognize their own status, even though we all 'know' the truth of the matter.

*'She's just not good enough to qualify as a barrister, but she's in denial about that.'*

*(In other words, although I think she doesn't have what it takes, she disagrees and is going to give it a go. She might turn out to be right.)*

Whenever you meet someone who refuses to admit they are wrong (and you are right), always accuse them of being in denial. It then ceases to be a matter of simple disagreement between you and them, and becomes instead their willful refusal to accept known facts.

# Denying the antecedent

As with *affirming the consequent,* the fallacy of *denying the antecedent* is for those who do not really care if their brain is going forwards or backwards. It does not admit the possibility that different events can produce similar outcomes.

*If I eat too much, I'll be ill. Since I have not eaten too much, I will not be ill.*

(So saying, he downed a whole bottle of whisky, cut his hand on a rusty nail and sat out all night in wet clothes.)

The point is, of course, that other events can bring about the same result, even if the event referred to does not take place. With these 'if . . . then' constructions, it is all right to affirm the antecedent (the 'if' part) and it is all right to deny the consequent (the 'then' part). It is the other two which are fallacious, affirming the consequent or denying the antecedent.

*If he's slow, he'll lose.*
*Since he isn't slow, he won't lose*

(But he might just be stupid.)

You can affirm the antecedent: 'He is slow; he will lose.' You can deny the consequent: 'He didn't lose, so he can't have been slow.' The first of these is a type of argument called the *modus ponens,* the second is called the *modus tollens,* and both are valid. It is the other two which are fallacies, even though they resemble the valid forms.

Denying the antecedent is a fallacy because it assigns only one cause to an event for which there might be several. It dismisses other possibilities which could occur.

The fallacy commonly occurs where plans are being laid. It engenders the belief that if those things are avoided which bring harmful consequences, then a pleasant outcome can be expected:

*If I smoke, drink or have sex, it will shorten my life-span. I shall give up cigars, booze and women and live another hundred years.*

(No. It will just *feel* like a hundred years.)

It occurs to no lesser degree on the international scale. Countries may calculate the courses of action which bring unpleasant consequences in their wake. What they are not able to do is insure themselves against even worse outcomes simply by avoiding those actions.

*If we have a strong army, countries which fear it might attack us. So by disarming, we remove that risk.*

(Possibly, but they might be more likely to attack because it brings no retaliation.)

You can use the fallacy of denying the antecedent very skilfully in support of the status quo. It is a natural conservative fallacy because most changes we make do not avert all of the evils of the world. By pointing to the likelihood that death and taxes will be the result of the proposed actions, you might lull an audience into rejecting them. The fact that death and taxes will result anyway should not impinge on your success.

# Dicto simpliciter

*Dicto simpliciter* is the fallacy of sweeping generalization. It consists of the application of a broad general rule to an individual case whose special features might make it exceptional. To insist that the generalization must apply to each and every case, regardless of individual differences, is to commit the fallacy of *dicto simpliciter.*

> *Of course you voted for the resolution. You're a dock-worker, and your union cast 120,000 votes in favour.*

(Carried unanimously, brothers, and by a clear majority.)

Many of our general statements are not universals. We make them in the full knowledge that there will be cases whose accidental features make them exceptions. We are apt to say that various things make people healthy, knowing that we do not necessarily have to mean 'all' people. We make similar generalizations about foods, even though we recognize that some people have allergies to various foodstuffs.

When we insist on treating a generalization as if it were a universal which admitted no exceptions, we commit a *dicto simpliciter.* The fallacy arises because we use information about the whole of a class, which has not been established or accepted. We bring in outside material, therefore, without justification.

> *Everyone knows that hooded teenagers are criminals. Since this hooded one isn't breaking any laws, he must be older than he looks.*

(Or maybe he's just having a day off.)

*Dicto simpliciter* arises whenever individuals are made to conform to group patterns. If they are treated in tight classes as 'teenagers', 'Frenchmen', or 'travelling salesmen', and are assumed to bear the characteristics of those classes, no opportunity is permitted for their individual qualities to emerge. There are political ideologies which attempt to treat people in precisely this way, treating them only as members of sub-groups in society and allowing them only representation through a group whose values they may not, in fact, share.

> *Look, you're a civil servant. Your representatives voted for this action because they know it will be good for the civil service. It must therefore be good for you.*

(He only imagined those lost wages.)

In discussing people of whom we have a little knowledge, we often use *dicto simpliciter* in the attempt to fix onto them the attributes of the groups they belong to. Knowing only that a neighbour is civil to us and drives a better car, we try to deduce things from the fact that he is a Catholic or a squash-player. Our assumption of ancillary properties may, in fact, be correct; the mistake is to suppose that it must be: 'We all know that children are smaller than their parents. Well, now that I'm 50 and Dad is 80, I've noticed that I'm quite a bit taller. Maybe he isn't my real father.'

*Dicto simpliciter* can be used to fit people into stereotypical moulds. Since they belong to the class of Frenchmen, ballet-dancers and horseriders, they must be great lovers, effeminate and bow-legged. You must appeal to generally accepted truisms in order to fill in details about individual cases which would otherwise be resisted.

You should as a parent use *dicto simpliciter* to trick your child into doing what you want instead of what they want:

> *Spinach is good for growing children. Eat it up.*

(But beware of the construction which says that 'all good children do such and such'. Your progeny might slip out of the group in question by recognizing themselves as bad.)

# Division

The *doppelgänger* of the fallacy of composition is that of division. When we attribute to the individuals in a group something which is only true of the group as a unit, we fall into the fallacy of division:

*Welsh-speakers are disappearing. Dafydd Williams is a Welsh-speaker, therefore Dafydd Williams is disappearing.*

(No such luck. Only the class of Welsh-speakers is disappearing, not the individuals who make it up.)

We commit the fallacy by sliding our adjectives to describe the whole over onto the individuals who comprise it:

*The Icelanders are the oldest nation on earth. This means that Björk must be older than other pop stars.*

(And before you go to her house, remember that the Icelandic people live surrounded by hot mud and active volcanoes.)

As with composition, the source of the error in the fallacy of division lies in the ambiguity of collective nouns. Both of these are a form of the fallacy of equivocation, in that it is the different meanings of the noun which upset the validity of the argument. It would only be valid if the same meaning were retained throughout. (The gospels are four in number. St Mark's is a gospel, so St Mark's is four in number.)

Division is often used fallaciously to confer upon an individual some of the prestige attached to the group or class to which he belongs.

*The French are tops at rugby; Marcel is French; obviously he must be tops at rugby.*

(But, since the French produce a lot of low-fat milk, Marcel probably has some other strange qualities.)

*California is a very wealthy state, so if he comes from there he must be worth quite a bit.*

We often commit the fallacy unconsciously, typecasting people according to the groups from which they emanate. This can work to their advantage: The teaching at Edinburgh University is brilliant; Johnson lectures there, so he must be really first-class, or to their disadvantage: Switzerland is a very passive nation, so I don't think we can expect too much initiative from our Swiss directors.

An entertaining version of the fallacy is called the fallacy of complex division, and assumes that subclasses of the whole share the same properties as the entire class. In this version, we meet the average British couple with their 2.2 children, out walking their 0.7 of a cat with a quarter of a dog. They have 1.15 cars, which they somehow manage to fit into only a third of a garage.

In the world of complex division, an expectant couple with two children are very nervous, because they know that every third child born is Chinese. In the real world, of course, it is different subclasses which produce the overall figures for the class as a whole. ('Test-pilots occasionally get killed, so I imagine that Flight-Lieutenant Robinson will get killed now and again himself.')

Division can be used to bring unearned credit upon yourself by virtue of your membership of meritorious classes:

*Let me settle this. We British have a longer experience of settling disputes than anyone else in the world.*

(Most of it acquired long before any of us were born.)

It can also be used to heap odium upon adversaries by pointing similarly to their involvement in groups which command no respect.

*My opponent comes from Glasgow, not a city noted for high intelligence.*

(If it were true, it would probably be because the bright ones had, like your opponent, come from it.)

# Emotional appeals

It would be a strange world if none of us were influenced by emotions. This influence steps over the boundary into the territory of logical fallacy, however, when it becomes the means of deciding the soundness of an argument. The emotions which influence our behaviour should not influence our judgement on questions of fact. While it might be appropriate to show pity to a convicted criminal, it is certainly not sound procedure to let pity affect our judgement of whether or not he committed the crime.

Recognition that reason and emotion have separate spheres of influence is as old as Plato's division of the soul. David Hume put it succinctly, telling us that passion moves us to act, whereas reason directs the course of those actions. Emotion, in other words, motivates us to do things, but reason enables us to calculate what to do.

Separate spheres they may inhabit, but sophists and tricksters have long known ways of making emotions invade the territory of reason. Once whipped up, the emotions can be set at such a gallop that they easily clear the gulf between their domain and that of reason. A complete range of fallacies is available, with as many names as there are emotions to draw on.

In addition to the ones important enough or common enough to be covered by separate treatment, there is a list of assorted and miscellaneous emotions, complete with Latin tags, which can be drawn upon at one time or another to lead reason astray from its intended course. The unwary sailor is entranced by the alluring calls of the appeal to fear (*argumentum ad metum*), to envy (*ad invidiam*), to hatred (*ad odium*), to superstition (*ad superstitionem*) and to pride (*ad superbiam*). There is even more to tempt our preference for a quiet time with an appeal to a just proportion of everything (*ad modum*), and one which actually says straight out that sentiment is a better guide than reason (*sentimens superior*). Unless one deliberately blocks out the pull of these emotions, as the sailors of Odysseus blocked up their ears to the allure of the Sirens' call, it is difficult not to be influenced. Therein lies their enduring effectiveness as fallacies.

> *Those who still oppose nuclear disarmament should study the effects of a thermonuclear blast. It can melt the eyeballs and vapourize human flesh from great distances.*

(This *argumentum ad metum* can be intensified by the use of photographs and films and simulated burns, and anything else which might distract the audience from asking whether nuclear disarmament will make it more or less likely.)

> *There is no way in which Robinson could have solved the problem. That would make him better than we are.*

(Right. Envy will not affect the outcome, though a timely *ad invidiam* might persuade people not to believe it.)

The secret of using these fallacies is a simple one. Take the trouble to discover the emotional disposition of your audience and use language calculated to arouse that emotion. When you have built it up assiduously by means of graphic descriptions, you turn it to bear on the question of fact. Very few audiences are able to turn it off abruptly; most will allow it to flood out onto the area normally reserved for reasoned assessment. Whether your appeal is to fear, envy, hatred, pride or superstition makes no difference. Indeed, you can use them alternately. Pride in one's own race, class or nation can be appealed to, even as envy of others is built up, perhaps to the point where an *ad odium* becomes possible.

The *argumentum ad modum* deserves a special mention because its appeal is to the audience's desire for gradualism. An audience is most vulnerable to it when they are trying to be reasonable. They equate reason with a quiet life, thinking that something admitted in due measure is more likely to be right. Like the *argumentum ad temperantiam,* which urges the middle course between extremes, the *ad modum* appeals to that most ancient of maxims which recommends moderation in all things. You should always introduce your subtle appeal to lure them away from reason by urging your audience:

*Let's be reasonable about this.*

(A strong emotional appeal for the quiet life.)

*Sentimens* is a clever fallacy. Its idiotic claim, that emotion is a better guide, is most alluring to an intelligent audience. Intelligent people are often afraid of being thought rather cold because they use reason so much. They do not want to appear to be emotionally deficient, and are easy prey to a speaker who assures them that they are just as sensitive, loving and compassionate as the next person, who is also a bit of a bore. This permits them the delusion that they are welcome into the common fold, instead of remaining aloof from it. They happily abandon reason as the price of their admission ticket to the human race.

An individual can be ensnared with *sentimens,* and led to drop a carefully thought-out position after being assured that he or she cares just as much about humanity as the rest of us. A denial would hardly count as a good response. A crowd is even easier to lead by the nose on a sentimental string. I have rarely seen an international gathering which did not give a standing ovation to any gaga idiot who urged them to forget reason and concentrate on loving each other.

'Most of the troubles of this world are caused by people thinking things out, instead of responding naturally with warmth and humanity. We should ignore these facts about Third World dictatorships, and reach out with love in our hearts and . . .'

(And yuk.)

# Equivocation

Equivocation means using words ambiguously. Often done with intent to deceive, it can even deceive the perpetrator. The fallacy of equivocation occurs when words are used with more than one meaning, even though the soundness of the reasoning requires the same use to be sustained throughout.

*Happiness is the end of life.*
*The end of life is death;*
*So happiness is death.*

(The form of the argument is valid, but 'the end of life' refers to its aim in the first line, and to its termination in the second. With this discovery, out go a million schoolboy conundrums.)

*Half a loaf is better than nothing.*
*Nothing is better than good health;*
*So half a loaf is better than good health.*

Equivocal use of words is fallacious because it invites us to transfer what we are prepared to accept about one concept onto another one which happens to have the same name. Logic, which processes the relationship between concepts, is useless if the concepts themselves change.

*Elephants are not found in Britain, so if you have one, don't lose it*
*or you will never find it again.*

(The word 'found' represents two different concepts here.)

Many of the equivocal uses are easy to spot. Many more of them are not. Clairvoyants specialize in equivocal expressions to give them cover in the event of quite different outcomes. Politics would be a totally different art if it had to forego the fallacy of equivocation. So would business correspondence:

*You can rest assured that your letter will receive the attention it fully deserves.*

(As it executes a gentle parabola towards the bin.)

'Anyone who gets Mr Smith to work for him will indeed be fortunate.'

Puns and music-hall jokes often depend on this fallacy.

*'My dog's got no nose.'*
*'How does he smell?'*
*'Terrible!'*

Calvin Coolidge was asked:

What do you think of the singer's execution?

(He replied: 'I'm all for it.')

The advice given to a political candidate facing a selection committee is 'When in doubt, equivocate.' The blunt fact is that you cannot please all of the people all of the time, but you can have a good shot at fooling most of them for much of it. The candidate assures those in favour of capital punishment that he wants 'realistic' penalties for murder. To those against, he wants 'humane consideration'. But he could be in favour of realistic light sentences or humane killing.

Equivocation is a particularly tough paste for pouring into the cracks of international discord. It joins irreconcilable differences with a smooth and undetectable finish. Many full and frank discussions are terminated happily by the appearance of a joint treaty whose wording is carefully chosen to mean entirely different things to each of the signatories.

The vocabulary of equivocation may be learned from the strangers' gallery of the House of Commons. If you have a seat in the chamber, there is nothing you have to learn about it.

Once you have acquired the knack, and are fluent in phrases such as 'having due regard for', you can move on to the more subtle manifestations of the fallacy.

*Well, it all depends on what you mean by full-hearted consent.*

(You might have thought it obvious. You'd be wrong.)

# Ethical superiority

It is not a fallacy to be ethically superior to your opponent. It is a fallacy to assume you are without supporting evidence. And the evidence must be more compelling than the fact that your opponent disagrees with you.

The reasoning goes roughly like this: Since I am moral and hold these views, anyone who holds alternative views must be immoral. But the fact that I am moral and have blue eyes doesn't make brown-eyed people wicked. People with differing views might be true to a different moral code, or they may share your moral code but disagree on the steps it takes to bring about a moral outcome. We might agree that everyone should be brought up to a minimum standard of living, but disagree as to whether this should be brought about by state handouts, private charity or higher wages, or by exempting them from the taxes that others have to pay. Supporters of one approach are not necessarily superior ethically to those who favour a different method.

*It is right to imprison those who refuse to pay the BBC television licence fee, and those who oppose this are simply encouraging lawlessness, selfishness and a lack of respect for decent social behaviour.*

Yes, but they just might think it immoral to force people to pay for things they don't watch, or perhaps think that television is more appropriately funded by non-compulsory methods of finance. The fact that they disagree with you does not make them wicked.

This fallacy has been called 'The Tin Man' fallacy (as opposed to 'The Straw Man') because of its presumption that the opponent, like the tin man in *The Wizard of Oz*, has no heart.

*'When there are homeless people on the streets, to oppose the proposed doubling of alcohol duties to help re-house them illustrates the very depths of hard-hearted depravity.'*

(You might care just as much but think there are better ways of doing it.)

This is an easy fallacy to use. All you have to do is to puff up with righteous indignation at those who oppose your views, and point to their reckless disregard for the poor, the deprived, the sick, the destitute and any other category you can think of. Your own policies on the other hand show your moral superiority by your preparedness to spend other people's money on everything you care about.

# Every schoolboy knows

You would be amazed what every schoolboy knows. Anxious to secure acquiescence in their controversial claims, disputants solemnly assure their audiences that every schoolboy knows the truth of what they are saying. The audience, not wishing to be ignorant of matters so widely understood by children, are supposed to keep silent about their doubts. Thus complex and dubious assertions are passed off unquestioned.

*Every schoolboy knows that the rate of gene loss from a closed reproductive system is expressed by a simple and well-known formula.*

(Indeed, this is the main topic of conversation over catapults and conkers.)

The tactic is fallacious. Its basic purpose is to appeal beyond the evidence to secure acceptance. The audience is invited to assent not from conviction but out of shame and fear of being thought less knowledgeable than a mere child. The merits of the point are meanwhile overlooked.

So widely used is the tactic that the hapless youth is now encumbered with several encyclopaedias of knowledge. There is scarcely anything which he does not know.

*As my learned colleague is doubtless aware, every schoolboy knows that it was* Rex v. Swanson *which established in 1749 the precedents governing the use of coaching horns on the public highway.*

(And you can be sure that the same gifted, if youthful, legal scholar is also aware of the judgment in *Higgins v. Matthews* 1807.)

The aforementioned schoolboy has an intuitive grasp of the obvious, and has been widely praised for this ability:

*Why, it is obvious even to a mere child that interstellar dust clouds would long ago have been excited to incandescence and be emitting black-body radiation were it not for the expansion of the universe.*

(It is not quite clear whether the mere child finds this obvious even before he becomes every schoolboy, or whether he picks it up after a few lessons.)

The fallacy is a special case of the more general fallacy of false advertisement, which consists of advance praising of your own views. Since you precede them by the information that they are known to every schoolboy and obvious to a mere child, you are scattering roses in their path. The fallacy may be perpetrated no less effectively by opening with 'Obviously' contentions which are by no means obvious.

*We hold these truths to be self-evident.*

(So anyone who doesn't agree must be really stupid.)

To use the fallacy effectively, you should never enter an argument without taking half the kindergarten class along for the walk. As well as the mere child and every schoolboy, you will need even a half-wit, albeit a very knowledgeable one. Every beginner should be in your posse to instruct the experts, and for sheer range of vision you will need everyone.

*'Everyone can see that . . .'*

(Even where no one else but you has such sharp eyes.)

When putting across a really controversial point, you might as well send the whole team into action:

*Every schoolboy knows the description of the visitors in Ezekiel; and even a half-wit realizes that ancient disasters were caused by cosmic disturbances. A mere child could work out that extraterrestrial forces are involved, so it is obvious to everyone that Earth has been under attack for centuries. Now as even beginners to the study of UFOs know only too well . . .*

(By this time your schoolchildren and half-wits should have cleared everyone else off the field.)

Do beware of actual schoolboys though. If there is one in your
audience, the smart alec is quite likely to step forward and contradict
you with the facts. Some of them are too good.

# The exception that proves the rule

Exceptions, of course, disprove rules. Despite this, many people confronted by a counter-example to their claim will dismiss it as 'the
exception that proves the rule'. The fallacy consists in the dismissal of
a valid objection to the argument.

> *'You never find songs written about any towns in Britain apart from
>     London.'*
> *'What about "Scarborough Fair?"'*
> *'That's the exception that proves the rule.'*

(If one leaves Liverpool and Old Durham Town out of it.)

The origin of the fallacy lies in the changing uses of language. The
word 'prove', which is now taken to refer to establishing something
beyond doubt, used to mean 'test'. Something would be 'proved'
to establish its quality, and this is the sense which has passed down
to us in this fallacy. The exception puts the rule to the test and, if it
is found to be a valid exception, refutes it instead of proving it in the
modern sense of the word:

> *No fictional character ever attracted fan clubs in distant countries
> like pop stars do. Sherlock Holmes does, of course, but he's simply
> the exception that proves the rule.*

(An elementary fallacy, dear Watson.)

There is a very loose way in which an exception can help to point to an
otherwise general truth. If we all recognize an exception as remarkable,

and identify it as such, then it does show that we accept that the rule which it counters does usually apply. In this sense, the one case we recognize as a freak points to the otherwise general truth:

*Medical advances are made by painstaking research, not by chance. I know there was penicillin, but everyone knows that was a chance in a million.*

(Whether true or not, this is a legitimate line of argument provided the rule is not claimed as universal. Everyone's acknowledgement of the unique exception points to a rule which says the opposite, with this one exception.)

Even in this specialized case, the exception disproves the universal rule. The trouble with sweeping statements is that it really does take only one exception to negate them. The medieval world abounded with universals which assured people that the sun would always rise and set each day, and that there could be no such thing as a black swan. A visit to the land of the midnight sun scuppered the first one, and the discovery of black swans in Australia polished off the second. It would be pleasant for many people if they could live in a world of certainties, surrounded by huge general truths. Exceptions come baying at that cosy world like wolves at the fringes of a camp-fire. They introduce uncertainties and doubts, and the temptation there is to use the fallacy quickly to get rid of them so that we can continue as before.

The exception that proves the rule is a fallacy beloved of those who are emphatic in their judgements. They have the world neatly divided into categories, and do not intend the irritant sand of exceptions to intrude into the well-oiled machinery of their worldview. In their smooth-running world, all pop stars are drug addicts, all feminists are lesbians and all young people are weirdos. Any honourable exceptions to the above categories are ejected with equal smoothness as 'exceptions that prove the rule'. The great thing about this particular fallacy is that it renders your argument invulnerable to factual correction. The most embarrassing proofs that you are just plain wrong can be swallowed whole as 'exceptions that prove the rule', and need occasion no more than a slight pause in your declamation.

> *'Lend us a fiver. I've always paid you back before.'*
> *'What about last week?'*
> *'That was the exception that proves the rule. You know you'll get it back in the long run.'*

(Put on your trainers.)

# Exclusive premises

The standard three-line argument called a syllogism has two premises and a conclusion; the premises are the evidence and the conclusion is deduced from them. If both of the premises are negative, no conclusion can be validly drawn from them and the fallacy is called the fallacy of exclusive premises.

> *No handymen are bakers, and no bakers are fishermen, so no handymen are fishermen.*

(It seems innocent enough, but the logic is fishier than the handymen. If we had used 'tax-dodgers' instead of fishermen, we would have ended up saying 'no handymen are tax-dodgers', which everyone knows is untrue. The fault lies with the two negative premises.)

The source of the fallacy is fairly clear. The three-liner relates two things to each other by means of the relationship which each has with a third. When both premises are negative, all we are told is that two things lie wholly or partly outside the class of a third thing. They could do this however they were related to each other, and so no conclusion about that relationship can be drawn:

*Some brewers are not idiots, and some idiots are not rich, so some brewers are not rich.*

(Did you ever hear of a poor one? With two negative statements, the idiots who are not rich do not need to be the same ones who don't include the brewers among their number. If this sounds confusing, remember two things: two negative premises do not prove anything, and all brewers are rich.)

The fallacy tends to occur because some people genuinely believe that if a group is excluded from something, and that group in turn is excluded from something else, then the first group is also excluded from it. If John cannot get into the Masons, and the Masons cannot get into the country club, it seems plausible to assume that John doesn't stand a chance for the country club. Of course, since the Masons cannot get in, John might stand a better chance because he isn't one of them.

*No pudding-eaters are thin, and some smokers are not pudding-eaters, so some smokers are thin.*

(Many of us are negative about puddings, but two negative statements about them will not tell us anything about smokers. If smokers are thin, it might well be from worrying about the health warnings, and a lack of money to eat puddings after paying for the cigarettes.)

When you want to use the fallacy of exclusive premises, you should try to make your negative statements ones which the audience will accept the truth of. When you slip in a conclusion which seems plausible, they will assume that you have proved it. You will not get very far if you start out with statements such as 'No council workers are lazy', but should try instead to keep within the bounds of your audience's experience. Use obvious truisms such as 'No removal men are careful.'

# The existential fallacy

It is a curious feature of logic that statements which refer to the whole of a class do not actually tell us whether there are any members of that class.

*All cats are selfish.*

(This tells us that *if* there are such things as cats, then they are selfish. It does not imply that there are cats, any more than the existence of unicorns could be deduced from a similar statement about them.)

Statements which tell us about some of a class, however, do imply the existence of members of the class.

*Some cats are selfish.*

(This tells us that there are such things as cats, and that some of them are selfish.)

The existential fallacy occurs when we draw a conclusion which implies existence from premises which do not imply that. If our premises are universals, telling us about 'all' or 'none', and our conclusion is a particular one telling us about 'some', we have committed the fallacy.

*All UFOs are spaceships, and all spaceships are extraterrestrial, so some UFOs are extraterrestrial.*

(This seems harmless enough, but it is not valid. We could have said *all* UFOs were extraterrestrial, but by limiting it to some we imply that they exist.)

It seems puzzling that we can be more entitled to say that all are, than to claim that only some are. We can console ourselves with the thought that perhaps we have to know some of them to start talking about the features which apply to some but not the others. The universal statements, by not sorting any out, carry no such implication.

The fallacy consists of putting into the conclusion something for which no evidence was offered, namely the presumption that what is being talked about actually exists. By going beyond the evidence, we enter into the territory of the fallacy.

> *All policemen are tall people, and no honest Welshmen are tall people, so some honest Welshmen are not policemen.*

> (Alas, no evidence has been produced to show that there is such a thing as an honest Welshman.)

A conclusion about *all* honest Welshmen would have been acceptable, because it would refer only to any who *might* exist.

The existential fallacy is clearly the domain of those who wish to engage in rational discourse about astral forces and demonic entities but who suffer from the minor disadvantage that there is no evidence that any of these things exist at all. Statements are made telling what the things must be like if they do exist, and somehow we begin to encounter claims made about some of them. At that point, unknown to the audience, the assumption of real existence has been slipped in without evidence, like an ace dropped furtively from the sleeve.

> *All psychic entities are affected by human emotions, but some of them are more sensitive than others, and tend to be aroused by fear and hatred.*

> (And the same is true of invisible frogs, spotted Saturnians and warm-hearted Swedes. Before you can start sorting them out, you must first catch your hare.)

Use of the existential fallacy is surprisingly easy. Most audiences will respect your modest claims if you move down from assertions about all things to claims made for only some of them. This readiness is the gate through which you can drive a coach and six loaded with fairies and hobbits, ectoplasm and elementals. The malleability of human nature and the perfectibility of man went through the same gate long ago.

# *Ex-post-facto* statistics

A statistician has been described as someone who draws a mathematically precise line from an unwarranted assumption to a foregone conclusion. It is not quite as bad as that, but there are innumerable statistical fallacies ready to trap the unwary and aid the unethical. The fallacy of *ex-post-facto* statistics is perpetrated when we apply probability laws to past events.

> I drew the ace of spades. It was only a 1 in 52 chance, but it came up.

> (The same applied to all the cards, but one had to come up.)

We cannot draw too many conclusions from the low 'probability' of certain past events. Something has to happen, and if the range of possibilities is large, the probability of each one occurring is small. Whichever one occurs thus has a low probability. The fallacy is committed when we go on to suppose, from the occurrence of events of low probability, that something supernatural was operating:

> *I met my aunt in Trafalgar Square on Wednesday. Think of the hundreds of thousands going through the square that day, and you'll realize how unlikely it was that we should meet there. Maybe we are telepathic.*

> (And the same applies to the thousands of others you met.)

The probability of heads coming up four times in a row is only 1 in 16. The same is true of every other combination which might come up; only one thing is certain, that a 1 in 16 chance *will* come up if you make four tosses. The fallacy goes beyond the evidence, using statistics in an inapplicable way to point to mysterious influences where none are needed. *Ex-post-facto* statistics often appear in speculations concerning the origins of life and the universe. Exotic calculations are trotted out showing the incredible unlikelihood that things could ever have happened as they did:

*How lucky we are that our planet has just the right temperature range for us, and just the right atmosphere for us to breathe. It has to be more than luck.*

(Ten-legged blue things breathing ammonia on the third planet of 70-Ophiuchi are even now saying the same thing.)

Similar claims are made of the probability of the right chemicals coming together to form life. The fact is that in our universe chemicals combine in certain ways. If they were different, no doubt different beings in a different universe would be congratulating themselves on their good fortune.

The fallacy is a great prop for those who suppose themselves the children of destiny. Looking at the unlikely events which have led to their present position, they see the unseen but inexorable hand of fate behind them, never realizing that had things been different they could have said the same.

*Just think, if we hadn't happened to be staying in the same hotel, we might never have met and never have married.*

(But they might have met and married other people, and thought themselves just as fortunate.)

Your use of this fallacy will depend a great deal on your temperament. It can be deployed at short range to convince others that you are a favoured child of the universe and entitled to receive special consideration:

> *I believe I was* meant *to get this job. I saw the advertisement for it in a paper the wind blew against my face in Oxford Street. I feel that something put me in that place at that time so that I would get this job. I'm not saying that should influence your decision, but . . .*

(But it should. Few of us like to confront the remorseless hand of destiny by stamping on its fingers.)

If you have the other temperament, you can always use the fallacy to gain some sympathy:

> *Just my luck! Of all the parking meters in London she could have been checking, it had to be mine. And just at the worst possible time!*

(If you can fit going to the pub in between being used as a punch-bag by the universe, lines like this should be good for the odd sympathetic pint.)

# Extensional pruning

We are guilty of extensional pruning if we use words in their commonly accepted meaning, but retreat when challenged into a strictly literal definition. The fallacy becomes possible because there are two ways of understanding what words mean. We can describe the properties of what we refer to, or we can give examples. The first is called the 'intension', and the second is the 'extension' of the word. We could convey the sense of an expression such as 'movie star', for example, either by describing the role of lead actors and actresses in films, or by listing several well-known stars.

Words carry nuances of meaning by their associations. Little tendrils of thought ripple around them, evoking all kinds of ideas dependent on past associations. These nuances are part of the meaning of the word, provided they are understood by user and hearer alike. The fallacy of extensional pruning takes place when the user subsequently retreats from that meaning by insisting upon only a literal 'intentional' definition.

*While I said I would accept an inquiry, I at no time said that it would be independent, that it would be a public one, or that its findings would be published.*

(He might be correct in a limited, technical definition of the word. But this is not what most people normally understand from the associations they make with previous inquiries.)

The fallacy is committed by saying one thing, but permitting another to be understood. A contention must be the same to both user and hearer, or no reasoned discussion is possible. There are two ways of committing this fallacy: one is to mislead at the outset, the other is to retreat to a restricted definition in order to escape weaknesses in the position.

*All we said was that we'd install a switchboard. We didn't say it would work.*

(Nor did they.)

Advertisers often take pruning shears over the extravagant claims they have made.

*We'll take your one-year-old car as trade-in, at whatever you paid for it.*

(Strictly speaking, you paid one sum for the car, and another sum for the tax. They are not offering to give you the tax back as well, whatever you might have thought.)

Friends who are free with advice often cut back the meaning in a similar way, after the consequences have emerged.

*Look, I know I said you'd feel like a millionaire. I know lots of millionaires who feel pretty miserable. Stop complaining.*

(You would feel like a swine if you hit him, but you probably know lots of swine who'd enjoy it.)

The extensional pruner announces his activity. Like the bow wave of an advancing ship, his utterances mark his passage. The inevitable 'all I said was . . .' and 'if you examine my exact words . . .' show him to be a man of great qualifications. You recognize him as the man who never really said at all what most people took him to be saying. The fine print one always watches for is in this case in the dictionary.

You can add extensional pruning to your repertoire once you are adept at making a limited statement pass itself off as a wider one. Gather yourself a collection of phrases whose meaning is understood by everyone, even though the words themselves are more restricted.

*I said I'd get you another drink if I was wrong: water is another drink.*

*I said I wouldn't have any more cigarettes until later in the week. Five minutes afterwards was later in the week.*

(Speak softly, and carry a big dictionary.)

# False conversion

False conversion takes place when we deduce from the fact that all cats are animals the additional information that all animals are cats. The converse of a statement, made by exchanging the subject and predicate, is true in some cases, false in others. When it is performed for one of the invalid cases, it is called false conversion.

*All rats are four-legged animals, so obviously all four-legged animals are rats.*

(This one is obviously false. Others are less so.)

*Some mortal beings are not cats, therefore some cats are not mortal beings.*

(It would be remarkable if the existence of beings other than cats were sufficient to establish the existence of an immortal strain of cats.)

The rule is intricate, but worth learning. We can make statements about all or some, and we can make positive or negative assertions. This gives us four types of statement:

**1** All are

**2** Some are

**3** None are

**4** Some are not

The rule is that only types 2 and 3 give valid conversions. If you exchange subject and predicate for types 1 and 4 you commit the fallacy of false conversion. The reason for the fallacy is that you cannot swap a distributed term (covering the whole of its class) for an undistributed one. In type 2, both subject and predicate cover only part of the class, and in type 3 they both cover all of it. Types 1 and 4 cannot be swapped around because they mix distributed with undistributed terms. What the rule means in practice is that you can swap around statements of the form

Some As are Bs and
No As are Bs

but you cannot swap those which tell us

All As are Bs or
Some As are not Bs

If we know that no innovative people are bureaucrats, we can deduce perfectly correctly that no bureaucrats are innovative people. What we cannot do is deduce from the knowledge that some journalists are not drunks the alternative statement that some drunks are not journalists. It may happen to be true, but we cannot deduce it from a false conversion.

In practice, most people can spot the obvious falsity of converting statements about all animals or all cats. The fallacy tends to be more common, and more deceptive, when it appears in the 'some are not' form.

*Since we know that some Marxists are not school-teachers, it follows that some school-teachers are not Marxists.*

(No it doesn't.)

Your own wilful use of this fallacy requires careful planning. It is a short-range tactical fallacy, and is best concealed by not letting the audience know if you are talking about 'some' or 'all'. The claim that 'Texas rabbits are animals which grow to more than a metre long' is skilfully ambiguous. It is not clear whether it refers to some Texas rabbits or all of them. Your surreptitious false conversion would then leave your audience convinced that any animal in Texas more than a metre in length must be a rabbit. It would also leave any Texans hopping mad.

# False precision

False precision is incurred when exact numbers are used for inexact notions. When straightforward statements about experience are decked out in numbers well beyond the accuracy of possible measurement, the precision is false and can mislead an audience into supposing that the information is more detailed than is really the case.

> *People say the Scots are mean, but they have been shown in surveys to be 63 per cent more generous than the Welsh.*

> (What measurement of generosity allows for that kind of a figure to be put on it?)

Both mathematics and science make widespread use of numbers, and both have prestige as sources of authority. The extension of exact numbers into areas quite inappropriate to them is often little more than the attempt to invest certain statements with the aura and prestige attaching to mathematics and science.

The fallacy derives from the use of unjustified material, and from the attempt to impart more confidence in the assertions than the evidence for them actually merits.

> *Our mouthwash is twice as good, yes two times as good, as its leading competitor.*

> (On what instrument, do you suppose, can one read off the quality of a mouthwash; and in what units?)

There are several versions of this fallacy, all of which have it in common that the numbers used give a misleading impression of the confidence one can place in the claim.

*Four out of five people can't tell margarine from butter.*

(It may be true, but how is it established? If large numbers on a one–against–one test repeatedly fail to distinguish them, we might be impressed. If smaller numbers fail to pick out the one margarine sample from a plateful of crackers covered with different types of butter, we might be rather less impressed.)

Yet another version might talk about quantity, where quality was a highly important factor.

*Kills 99 per cent of all household germs.*

(A worthy claim, unless the rest happen to be typhoid.)

False precision is as necessary to the continued happiness of many academics as are public money and whisky. Whole departments float upon it, just as some do on the other two ingredients. Those who are engaged in the study of human beings, for example, find few measuring rods scattered about. Because the real qualities of people cannot be measured, indices are constructed which can be measured, and then the indices are passed off as the real thing.

*Birmingham children are more racist than their London counterparts. A study of essays written by 10-year-olds showed that the London group used 15 per cent less racial epithets than their peer group in Birmingham.*

(The assumptions here are manifold. Maybe one can identify and agree upon what constitutes a racial epithet. Maybe their appearance in essays reflects their importance in the lives of children.

Maybe the use of them by children is evidence of racism. Maybe the cultural differences between Birmingham and London are not important etc., etc. None of these doubts qualify the opening line.)

Macroeconomists happily report that growth-rates were only 1.4 per cent, instead of the predicted 1.7 per cent, without telling us that some measurements of GDP cannot be taken within 5 per cent accuracy. Some figures for growth of GDP could be out by up to 10 per cent.

Psychologists measure the ability of children to solve set problems and call their answers intelligence. Social scientists measure how people respond to questions and describe the answers as a measurement of attitudes. False precision is like a hastily erected and flimsy bridge to carry our knowledge over from reality into the world of our desire. The load is more than it will bear.

Always use the fallacy when you need more authority for your claims. Behind the figures you quote, your audience will conjure up an army of white-coated scientists with horn-rimmed spectacles, and dedicated doctors with stethoscopes draped in careless urgency. The invisible army will nod sagely in support of each precise statement you make, and if the audience might have doubted you they will be reassured by the phantom legions who underwrite your figures.

*Whatever the academic merits, single-stream schooling certainly produces more balanced children. Surveys have revealed 43 per cent fewer psychological abnormalities among groups which . . .*

(Just don't tell them that the abnormalities included self-esteem, competitiveness and the desire to learn.)

Remember to be exact, especially when you are being vague.

*We can be 90 per cent certain that Bloggs is the guilty man.*

(And 100 per cent certain that you cannot prove it.)

# The false zero sum game

A zero sum game is not itself a fallacy. It describes a situation in which the supply of something is limited, so the share enjoyed by one person affects that available for others. Bluntly put, it means that if one person obtains more, others must receive less. It describes a pizza. Take a larger slice yourself, and someone else must make do with a smaller one. The fallacy is committed when someone declares or assumes that there is a limited supply of an item when the reality is that more of it can be produced.

The fallacy is surprisingly common. Many people think that an act of exchange or a trade is a zero sum game. They assume that the value in the deal is fixed, and that one participant can only obtain more value at the expense of the other. They speak of 'getting the better of the bargain' or of 'coming off best.' In fact a deal takes place because each participant assigns a greater value to what the other has over that which he or she already has. By trading, each of them gains greater value than they had. Instead of one being the loser in a zero sum game, both are now wealthier than before.

*"No, you may not take home a bucket of sand from the beach. There might be none left for other children."*

(This would only be true if there were a fixed supply of sand on the world's beaches without the possibility of more being made, and that enough children with enough buckets could deplete it. In reality, more sand is made when tides grind rocks together.)

*Poorer countries can only become richer if the world's wealth is shared out more equally.*

(No, this assumes that wealth is in fixed supply, whereas it is not. The world's wealth has been increasing for centuries and still is. There is a great deal more of it than there was at the time of Alfred the Great. Poorer countries might become richer by using trade and exchange to create wealth, like the richer ones did.)

*'If I buy two televisions that means someone, somewhere, will be without one.'*

(Only if no-one makes any more televisions. We can dream, but it's not likely.)

If you use this fallacy, you will gain a popular hearing because it appeals to people's envy. When you point to the possessions of those better off, you can assure your audience that those with more goodies have made you all worse off in the process of acquiring them. With a certain degree of flair, you could use the fallacy to make pillage and looting look morally respectable.

# The gambler's fallacy

Few fallacies are more persistent in gambling circles than the belief that the next toss (or spin, or draw) will somehow be influenced by the last one. Gamblers, and others, are led into this fallacy by confusing the odds against a whole sequence with the odds against any event in that sequence.

The odds against a tossed coin coming down heads five times in a row are easy to calculate. The answer is

$$1/2 \times 1/2 \times 1/2 \times 1/2 \times 1/2, \text{ or 1 in 32}$$

If the first four tosses, despite the odds, come down heads, the chance of the fifth toss being heads is not 1 in 32, but 1 in 2, as it was for each of the other tosses. The previous tosses do not affect the odds for the next one. In random or chance events, each go is separate from previous or future ones. Most casual gamblers, seeing four heads in a row, would bet on tails for the fifth toss because five in a row is unlikely. The professional gambler would probably bet on heads again, suspecting a crooked coin.

> *Red has come up 13 times out of the last 20.*
> *That means we are due for a run of blacks. I am betting on black.*

(If the table is honest, the odds on black remain, as before, the same as the odds on red.)

There is widespread belief in everyday life that luck will somehow even out. The phrase 'third time lucky' is indicative of a general feeling that after two failures the odds for success improve. Not so. If the events are genuinely random, there is no reason for supposing that two losses improve the chances of a win. If, as is more common, the results reflect on the character and competence of the performer, the two losses begin to establish a basis for judgement.

> *I'm backing Hillary Clinton on this one. She can't be wrong all the time.*

(Oh yes she can.)

One area where previous events do influence subsequent ones is in the draw of cards from a limited pack. Obviously, if one ace is drawn from a pack of 52 cards containing four aces, the chances of another ace being drawn are correspondingly reduced. Professional gamblers are very good at remembering which cards have been drawn already, and how this bears upon forthcoming draws. Still other gamblers are very good at making up from their sleeves what the laws of chance and probability have denied them from the deck.

Many so-called 'systems' of gambling are based on the gambler's fallacy. If betting on a 1 in 2 chance, you double the stake after each loss, then when you do win you will recover your losses and make a modest gain. The trouble with this is that the rules of maximum stake, if not your own resources, will soon stop you doubling up. (Try the trick of doubling up ears of wheat on each square of a chessboard, and see how quickly you reach the world's total harvest.) Furthermore, the odds are that the sequence which it takes to beat such a system will occur with a frequency sufficient to wipe out all of the winnings you made waiting for it. Only one rule is worth betting on: the house always wins.

You can use the gambler's fallacy by appealing to a quite unfounded general belief that the universe is somehow fair.

*My argument for avoiding the west of Scotland is that it has rained there on about half the summers this century. Since it was fine for the last two years, the odds are that it will rain this year.*

(Things change, even in the west of Scotland.)

You may find the gambler's fallacy particularly useful in persuading people to go along with you, despite a previous record which indicates that luck was not involved.

*I propose this candidate for our new secretary. I know that the last three I chose were pretty useless, but that's all the more reason to suppose I've had my share of the bad luck and will be right this time.*

(This sounds like bad judgement disguised as bad luck. The odds are that the new choice will be both pretty and useless.)

*The last four lawyers I had dealings with were all crooks. Surely this one must be better.*

(No chance.)

# The genetic fallacy

The genetic fallacy has nothing to do with Darwin or Mendel, but a great deal to do with not liking where an argument comes from. People give less credence to views which emanate from those they detest, regardless of the actual merit of the views themselves. Every time you dismiss an argument or opinion because you dislike its source, you commit the genetic fallacy. The fallacy is sometimes called 'damning the origin', and we can take it that the argument is sent to hell along with its source.

> *Don't be obsessed with punctuality. It was Mussolini who wanted the trains to run on time.*

(Mussolini's views on trains, whatever they were, are hardly an argument on punctuality. Bad men, especially verbose ones, are almost bound to say something right occasionally, much as a chimpanzee typing at random might produce *Hamlet.* No doubt Hitler favoured road safety and disapproved of cancer. Mussolini might have hit it lucky on the subject of trains.)

The genetic fallacy makes the mistake of supposing that the source of an argument affects its validity. Utterly wicked people sometimes utter worthy arguments, while saints are not immune from silliness. The argument stands alone, drawing neither strength nor weakness from its source.

This particular fallacy is often found basking in the hothouse world of fashionable ideas. A view from a currently fashionable source is given credence, but the same view would be rejected if it emanated from someone less modish.

> *The objections to the Council's new bus timetable come only from private property developers, and can be ignored.*

(Why? Private developers might well have legitimate opinions or insights on such matters. They are, alas, still *bêtes noires* in the world of local politics. Had the same objections come from Friends of the Earth they might have found more sympathetic ears.)

The genetic fallacy is nowhere more widely seen than in connection with the alleged views of a few universally detested figures. The association of Adolf Hitler with a viewpoint is generally sufficient to damn it. His predecessors, Genghis Khan and Attila the Hun, left fewer writings, but have many views attributed to them. In rare cases the hated name becomes adjectival, with the simple epithet Machiavellian or Hitlerian being sufficient to remove an idea from consideration by decent people.

*Tinkering with genes is fascist talk. That's what Hitler tried to do.*

(Actually, he did favour breeding from what he saw as superior stock, which is not necessarily the same as trying to eliminate certain disorders by gene-splicing. In view of his known association, one is surprised that the bloodstock industry and dog-breeding have gone as far as they have. For that matter, Volkswagens and autobahns seem to have caught on quite well, too.)

To use the genetic fallacy with devastating effect, all you need do is point out that your opponent is echoing arguments first put forward in Nazi Germany, then subsequently taken up by Augusto Pinochet and Saddam Hussein. You, on the other hand, are advocating points of view put forward by Mother Teresa, Princess Diana and Mary Poppins . . .

# Half-concealed qualification

In a half-concealed qualification, the words themselves express a limited claim, but the stress and construction is such that the qualifications are glossed over. Although the limits are stated, the audience barely notice them on the way to a discussion about a more general statement.

*Practically every single case of monetary expansion is followed within 16 months by an attendant general price rise of the same proportions.*

(This is the classic statement of sado-monetarism. Note the first word – nobody else does.)

In this example the qualifying word 'practically' is half-concealed by the stress given to *'every single case'*. Should embarrassing cases turn up which do not follow the rule, one can always retreat to the qualification and point out that the assertion did not claim to cover *all* instances.

There is a fallacy inherent in making a restricted claim and then engaging in discussion as if it were a general claim. The important information that the assertion does not apply to all cases is omitted from consideration. The fact that the limitation is expressed does not remove the fallacy. It is the fact that the qualification is half-concealed which causes it to be unnoticed, and which excludes relevant information.

*The link between poltergeist phenomena and psychological troubles is now clearly established. In almost every case of unexplained breakages and moving objects, there is a disturbed youngster in the household.*

(And since no one noticed the 'almost', we don't need to talk about the other cases.)

Half-concealed qualification is widely used to support half a case. When there is a gap in the evidence supporting a complete link, the fallacy papers over the crack. Science and philosophy do not admit unexplained exceptions. Newton would not have got very far by telling us that objects are *usually* attracted towards each other by a force which varies inversely with the square of the distance between them.

In daily life, however, we are less rigorous, and the fallacy finds room to make a partial case seem like a complete one.

*Palm trees don't normally grow in England, so it must be something else.*

(Normally he'd be right; but there are exceptions.)

Social engineering is often proposed on the basis of incomplete assertions governing how humans generally behave.

*Most crime is caused by juveniles, and nearly all young offenders come from broken homes. The answer to rising crime is not more police, but more family counselling centres.*

(Maybe it is. Let's hope the staff have as many qualifications as the argument does.)

There is a common character trait which will help you to get away with half-concealed qualifications. It expresses itself in a readiness to think of cases which do fit, rather than of cases which do not. On being given a limited statement, such as 'Most bosses flirt with their secretaries', many people will find themselves thinking of cases which they have known. Few find their thoughts led immediately to bosses who do not do this. You can use this propensity to have more read into your assertions than they are really claiming.

*Just about every Cambridge man working in the Foreign Office or security services in the late 1940s has turned out to be a spy and a traitor. Why don't we cut our losses, fire the rest, and not hire any more?*

('Just about every' seems in this case to mean a handful, or maybe three; but everyone will think of the ones they have read about who were exposed, rather than about the others who were not.)

# Hedging

Hedging is put around arguments, as it is around fields of crops, to prevent them from being trampled. Hedging in argument means sheltering behind ambiguous meanings so that the sense can be changed later. ('I said the last thing we wanted in the Middle East was an all-out war, and I stand by that. What we have embarked upon is a *limited* war . . .')

Hedging involves the advance preparation for a definitional retreat. The words and phrases are so carefully chosen that the option is retained to do a switch in definitions. Opposing arguments and examples bearing down on the arguer suddenly find a hedge barring their advance, while their quarry may be sighted in a different field. ('All I said was that I'd be home at a reasonable hour. I think that three o'clock in the morning is a reasonable hour in view of what I've been doing.')

Hedging is fallacious because it puts forward two or more different statements under the guise of one. The alternative interpretations are smuggled, like the companions of Odysseus, clinging to the undersides of the sheep which they appear to be. The hope is that the hearer, like the blinded Cyclops, will not know the difference. The effect of hedging is to render useless the information it purports to convey.

Soothsayers would be sorrier souls without hedging to give them more than one chance. Just as you hedge bets in a race by backing more than one horse, so in prophecy you can bet on more than one outcome.

> *Be bloody, bold and resolute; laugh to scorn / The power of man, for none of woman born / Shall harm Macbeth.**

(The hedge was that the witches failed to tell Macbeth that this description did not apply to those such as Macduff, born by

---

*William Shakespeare, *Macbeth*, iv, i, 79–81.

Caesarian section. He found this out after a very large hedge had moved from Birnam to Dunsinane.)

Most oracles and insurance agents are notorious for their use of hedging: some take it to unimagined lengths. The centuries of Nostradamus are so obscure, and can be translated in so many ways, that they can be used to predict literally anything. People have claimed to find in them the most astonishingly detailed, and astonishingly accurate, foretelling of the future. Not only Napoleon and Hitler, but even recent popes and politicians emerge from his pages. As with all hedged prophecies, however, there are tell-tale signals. People are very good at finding references in the writings of Nostradamus to what has already happened. They are not successful at finding accurate accounts of what will happen. There is also a remarkable consistency to the way in which subsequent ages have found that many of his prophecies made sense for their own time.

Dishonesty is an essential aspect of hedging. The ambiguity is inserted deliberately with intent to deceive, and for the purpose of proving the perpetrator correct, whatever the outcome. The fairground fortune-teller shelters harmlessly behind her hedge by telling you that you are destined to travel (even if only on the No. 36 bus home). The economist hides rather more wilfully behind the hedge that things will get worse, barring a major change in the world economy; (when they get better, it is because there was a major change in the world economy.)

Hedging requires planning. Few people can toss off ambiguous phrases on the spur of the moment; we expect to find them in the prepared statement which is issued, rather than in the off-the-cuff remark. You should accumulate a stock of phrases which look plain enough from one angle, but are bedecked with hedges as you approach them.

*You will find the cheque paid directly into your bank account.*

(When?)

# *Hominem* (abusive), *argumentum ad*

If you cannot attack the argument, attack the arguer. While an insult itself is not fallacious, it is if made in a way calculated to undermine an opponent's argument, and to encourage an audience to give it less weight than it merits. When this is done, the famous *argumentum ad hominem* abusive is committed.

> *Dr Green argues very plausibly for fluoridation. What he does not tell us is that he is the same Dr Green who ten years ago published articles in favour of both euthanasia and infanticide.*

> (Unless his argument is that fluoride will kill off the old people and infants more effectively, it is hard to see how this bears on the arguments for or against fluoride.)

The fallacy here, as with most fallacies of relevance, is that the argument is not treated on its merit. Arguments should stand or fall by their own qualities. Strictly speaking, the merits of the arguer do not come into it. Even the public relations industry is not always in error. It is only because we are reluctant to suppose that a good and sensible argument can come from a bad and stupid source that the *ad hominem* abusive has any effect.

> *Now I come to Professor Robinson's argument in favour of amalgamating the two colleges. Far be it from me to reopen old wounds by referring to the Professor's conviction three years ago for drunk driving, but we have to ask ourselves . . .*

> (Note the ritual denial. It is usually the signal for an *ad hominem* abusive, 'I don't wish to be catty, but – *miaow.'*)

There are many forms of this fallacy, some so specialized that they are identified and named as separate fallacies. Effective use demands a bold attempt to make the abuse appear to have some bearing on the issue under consideration. The use of personal attacks to cast doubt on the arguer's judgement gives one possible avenue.

Lawyers when cross-examining hostile witnesses tread a fine line between 'establishing the character of a witness' and a simple *ad hominem* abusive to discredit the testimony. Similarly, the use of witnesses on the character of the accused can often venture over the line into the territory of the fallacy.

The political arena is fertile territory in which some fallacies grow like weeds and others like carefully cultivated blossoms. The *ad hominem* abusive is one of the staples of parliamentary question-time.

*I would remind the House that when my questioner was in office unemployment and inflation doubled, and wages went down almost as fast as prices went up. And he has the temerity to ask me about the future of the mining industry.*

(No comment, which is what he is saying in a more circumlocutious form.)

Some of the poor quality of parliamentary debate can be laid at the door of the press. So long as there are sycophantic journalists prepared to praise an ordinary *ad hominem* abusive as a 'splendid riposte' there will be politicians labouring through the midnight hours to compose such gems as 'like being savaged by a dead sheep'. They perform to their audience.

The rules to remember when committing this fallacy are that the hostile material should, wherever possible, be introduced with apparent reluctance, and it should be made to bear on the question of whether your opponent deserves consideration by such a worthy and serious audience as you are both addressing.

*It is with a heavy heart that I release copies of these photographs and letters. I ask you whether this council can be seen to be influenced in its policy toward the new suspension bridge by a man whose behaviour with an 11-year-old girl flouts every standard of public and private behaviour which we, as a council, have a sacred duty to uphold.*

(Look out below.)

# *Hominem* (circumstantial), *argumentum ad*

In the *argumentum ad hominem* circumstantial, the appeal is to the special circumstances of the person with whom one is arguing. Instead of trying to prove the contention true or false on the evidence, its acceptance is urged because of the position and interests of those appealed to.

> *You can't accept the legitimacy of lending for profit. You are a Christian, and Christ drove the money-lenders from the temple.*

(This is not a general argument. It might not do much for a Hindu or a Jew, for example. The listener is invited to assent because of his Christian convictions.)

In a similar way people can be asked to accept a view because of their circumstances as members of the political party which supports it. In this version of the fallacy the error comes in by bringing the particular position of the audience into what is urged as a generally accepted truth. While such tactics might indeed convince that specific audience, they would not establish the rightness or wrongness of what is urged, nor the truth or falsity of a statement.

> *No one in this university audience can be opposed to handing out state money to subsidize services, otherwise you would not be here, occupying a subsidized place.*

(Actually, it is *other* state handouts which students oppose.)

A variant of the fallacy dismisses a person's views as representing *only* their special circumstances. It assumes that an oil company executive can reflect only his corporation's interest when he voices an opinion on the future of energy supplies. In the first place, the executive may well have independent views which differ from those of his company. In the second place, there is nothing to say that the corporation view is not the correct one, self-interested though it may be. The fallacy arises in this version by the wanton dismissal of possibly relevant material as much as by bringing in irrelevant matters such as the circumstances of the audience. Even if it can be shown why an opponent thinks as he does, it still does not show him to be wrong. ('As an opera-lover, you will be the first to agree that we need more subsidy for the arts.')

The appeal to special circumstances occurs in arguments addressed to specialist audiences. The American expression 'building a constituency' refers quite often to the process of adding together enough interest groups, all of which give support on account of their special circumstances. An adept, if unscrupulous, politician might build a power-base by directing argument not to the general good of society but to the special circumstances of public-sector employees, trade unions, welfare recipients, ethnic minorities and groups involved in sexual politics. The rightness or wrongness of the programme need not come into it if enough special circumstances can be appealed to.

Both versions of the *argumentum ad hominem* circumstantial can be used to advantage. You should employ the first version with respect to circumstances which are broad enough to include fairly large audiences. ('You, as members of the working class, will appreciate . . .') Especially useful to you will be the nominal membership of the Christian Church. Many people like to think of themselves as Christians, although they do not like the obligations which serious Christianity would impose. Thus when you appeal to them that as Christians they can hardly oppose your views, they will be forced into a reluctant and resentful acquiescence you could never have gained otherwise.

The second version is spectacular in the rejection of expert evidence against you. An expert is someone in the field, and as such his views represent only his circumstances as one who is involved. Thus, when the town-planner refutes your claims on town-planning; when the oil company expert shows what nonsense you have uttered on energy; and when the businessman exposes your cockeyed views about business, you smile sweetly in each case and observe: 'Well he would say that, wouldn't he?'

# Ignorantiam, argumentum ad

Socrates was thought by the oracle to be the wisest man because he alone knew how ignorant he was. The knowledge of ignorance might have been good at keeping Socrates modest, but it forms a poor basis for deduction. The *argumentum ad ignorantiam* is committed when we use our lack of knowledge about something in order to infer that its opposite is the case.

> *Ghosts exist all right. Research teams have spent many years and millions of pounds attempting to prove that they don't; and they have never succeeded.*

(The same could probably be said of Aladdin's lamp and the prospects for world peace.)

The positive version of *ad ignorantiam* asserts that what has not been disproved must happen. There is, in addition, a negative form which claims that what has not been proved cannot occur.

> *Talk of extraterrestrial life-forms is nonsense. We know there are none because every single attempt to establish their existence has failed utterly.*

(Also true of the Yeti, Bigfoot, the Loch Ness monster and editorial integrity.)

In both versions of the fallacy, the appeal is to ignorance. It is called upon to supply support for an assertion, even though our own state of knowledge does not normally bear on the truth or falsity of that statement. The fallacy consists of the intervention of irrelevant material, in the shape of our own ignorance, into an argument which is about something else. It is notoriously difficult to prove that something exists, especially if it is a shy creature which hides coyly in the deep of a Scottish loch, on the slopes of a mountain wilderness, or in the mists of the third planet of 61-Cygni. You practically have to meet one. Even then, a wealth of recorded evidence would be required to convince others.

To establish non-existence is even more difficult. You have to look at the whole universe simultaneously to make sure that your quarry is not lurking in any part of it. Not surprisingly, this feat is rarely accomplished, and thus leaves us with boundless spaces thickly populated with *ad ignorantiams* and the other products of our imagination.

*Kid, I've flown from one side of the galaxy to the other. I've seen a lot of strange stuff, but I've never seen anything that could make me believe there's one all-powerful force controlling everything.*

Of course there are cases in which our lack of knowledge does influence our judgement; they occur where we would expect to have that knowledge if the thing were true. One would rightly reject a report that Camden Town Hall had been swallowed whole by a slime monster if there were no reports of it in the newspapers, no eyewitness accounts on television, no street celebrations, or any of the evidence we would expect to accompany such an event.

The *ad ignorantiam* forms the semblance of a cloak to cover the otherwise naked beliefs of those who are predisposed to give credence to extraordinary things. Under its comforting warmth shelters a widespread popular belief in telepathy, poltergeists, demonic possession, magic pyramids, Bermuda triangles and the innocence of tobacco. ('Television violence doesn't do any harm. None of the surveys has ever managed to prove that it does.')

The *argumentum ad ignorantiam* is useful if your own views do not follow received opinion. You can persuade others to share these bizarre notions by appealing to the lack of evidence to the contrary. Only slight difficulty is occasioned by the abundance of evidence in many cases to prove you wrong: you reject the evidence, deploying further *ad ignorantiams* to show that no one has ever proved the evidence to be reliable. In this way you will be able to sustain a preconceived view of things in the teeth of all sense and experience. When you are expert at it, you can add the letters *'ad ign.'* after those denoting your degree in sociology. After all, no one can prove that you shouldn't.

# *Ignoratio elenchi*

*Ignoratio elenchi* is one of the oldest fallacies known to us, being first identified by Aristotle. When someone believes himself to be proving one thing, but succeeds in proving something else instead, he commits *ignoratio elenchi*. He not only argues beside the point, but directly to a different conclusion.

> *I shall oppose this measure to permit people to leave school earlier by proving once again the value of education.*

> (Proving the value of education does not prove the case against permitting earlier leaving. Perhaps it takes education, as opposed to schooling, to see the difference.)

The thesis which is proved is not relevant to the one which the arguer sought to prove, which is why this is sometimes known as the fallacy of irrelevant thesis. The fallacy consists of supposing that the one conclusion equates with the other, when in fact they make separate points. The arguments which would support the first conclusion are omitted, and those which support the irrelevant conclusion are brought in instead.

> How could my client have ordered the murder? I have proved beyond a shadow of doubt that he was not even in the country at the time.

> (Well done. Does that show he didn't order it before he left, or arrange it by telephone?)

*Ignoratio elenchi* has a subtle appeal. Its strength lies in the fact that a conclusion is validly proved, even though it is the wrong one. Anyone who concentrates on the argument may well find that its soundness diverts his attention away from the irrelevant conclusion.

> Is gambling a worthwhile occupation? Believe me, we not only work as hard as anyone else, but harder. It takes hours of study every day, quite apart from the time spent doing it.

> (OK, it's hard work. Now, is it worthwhile?)

*Ignoratio elenchi* makes its brief, but usually successful, appearance wherever someone accused of doing something he did do is quite prepared to deny something else. It is a central feature of all points where journalistic and political circles touch. The use of the fallacy has an almost ritual quality to it. Whether it is under the steady beam of studio lights, or the staccato illumination of flash cameras in the streets, the little tableau is enacted. The eager pressmen solemnly charge the great man with one thing, and he, with equal solemnity, shows that he has not done another.

> *'Isn't it true, Minister, that you have allowed the living standards of the poor to fall in real terms?'*
> *'What we have done is to increase by 3.7 per cent the allowance to childless dependent females, and by 3.9 per cent the allowance to*

*widows with two children, these increases both being larger than our*
*opponents ever managed in a single year of their term of office.'*

In the more relaxed atmosphere of a studio interview, the great man
will often brazen it out, with royal trumpeters announcing his *ignoratio*
*elenchi*:

*Well, John, that's not really the point, is it? What we have done*
*is to . . .*

(And you can bet that this *certainly* isn't the point.)

Obviously you can use the fallacy for close-quarters defensive work.
Your audience will be so impressed by all the things which you can
prove you have not done, that their attention might wander away from
those you have. The more laborious and detailed your proofs, the less
chance there is of anyone remembering what it was you were actually
accused of.

You can also use it in an attacking role, proving all kinds of things
except the ones that matter. There are many things which can be
demonstrated about nuclear power, hunting animals and refined white
sugar which are not relevant to the central topic of whether others
should be banned from doing things you do not approve of.

*Jogging in public should be banned. There are studies which show*
*it can increase the risks to health, rather than decrease them.*

(Even if it were true, would it be an argument for banning public
jogging? It sounds as though the major adverse effect is not on the
jogger's health, but the speaker's conscience.)

# Illicit process

There is a rule about arguments which tells us that if a term in the
conclusion refers to the whole of its class, then the evidence pointing

to that conclusion must also have told us about the whole class. We cannot reach a conclusion about 'all estate agents', for example, unless we start with some knowledge which applies to all of them. To know that some estate agents are guilty of this or that practice will not justify us reaching conclusions about all of them. Arguments which break this rule are said to commit the fallacy of illicit process.

*All tax-collectors are civil servants, and all tax-collectors are bullies, so all civil servants are bullies.*

(Too harsh. There may be some somewhere who are just a little overbearing. The fallacy is that we refer to *all* civil servants in the conclusion, but the premise only tells us that tax-collectors are *some* of them.)

The argument which uses illicit process has to be fallacious because it makes unsupported claims. Although the premises talk only about some of a class, the conclusion introduces for the first time the rest of that class. In other words, we try to reach conclusions about things we have no evidence on, and commit a fallacy by doing so.

There is another version of illicit process which is harder to spot:

*All cyclists are economical people, and no farmers are cyclists, so no farmers are economical people.*

(This appears to fit the observed facts, but there is a fallacy. We could just as easily have said 'All cyclists are mortals'. This would give the distinct impression that big fat farmers would be driving their big fat cars for ever.)

The source of the fallacy in this example is that the premise tells us that cyclists are some of the class of economical people. The conclusion, on the other hand, tells us that the *entire* class has not a single farmer in it. Again, the fallacy is illicit process.

These terms which cover the whole of their class are called 'distributed terms', and there is a rule for finding them. Universals, which talk about 'all' or 'none', have distributed subjects; negatives, which tell us what is not the case, have distributed predicates. In the example above, the term 'economical people' is distributed in the conclusion, since it is the predicate of a negative statement. In the premise, however, it is undistributed, being neither the subject of a universal nor the predicate of a negative. It sounds complicated, but the rule makes it simple. You will soon be seeing which conclusions try to cover all of a class without any information to justify it. To dazzle your friends totally, you should call the fallacy illicit minor when the subject of the conclusion is unjustifiably distributed, and illicit major when the predicate of the conclusion is so treated.

To use illicit process requires a good deal of homework. You should deploy it in support of conclusions which look plausible but have the minor technical drawback that you cannot prove them. Your expertise at illicit process will enable you to construct arguments based on what some of the class do, and slide smoothly into conclusions about all of them.

*Some Australians are pleasant fellows, and some con-men are not pleasant fellows, so some Australians are not conmen.*

(Who knows? It may even be true; but it takes a lot more than this to prove it.)

# Insulation from alternatives

Arguments and ideas are supposed to live in a rough old world, constantly exposed to attack from contradictory ideas and arguments. They should face, and deal with, hostile evidence that might undermine them if it cannot itself be undermined, explained or refuted. This is how J S Mill thought that truth would emerge, by success in open conflict. It is how science progresses, with theories being exposed to experiments that produce incompatible results, and with counter theories that explain the evidence with different assumptions. A 'scientist' who kept all his or her theories secret, except from colleagues who agreed with them, and did not publish the results of experiments, would be regarded as undeserving of the name, and rightly so, since open debate is part of the activity.

In the world of intellectual debate there are those who surround themselves only with like-minded people, who take information only from like-minded sources and who never expose their ideas to the cut and thrust of genuine opposition. The problem with this approach is that the person who practises this comes to regard the ideas shared among the like-minded clique as the only valid ideas. Never having exposed them to real hostility, they think the views they hold represent the truth, or at least the only reasonable account of the way things are.

There is a parallel with isolated communities cut off from contact with the wider world. They do not regard their practices just as their way of doing things, but as *the* way of doing things. When they do encounter a foreigner who lives differently, they are usually shocked and appalled by the person's lack of decency. The point is that you need contrasts to make comparisons and evaluations, and ideas that face no competitors lack a way of proving their worth.

If people work in a broadcasting organization and are surrounded only by those holding similar views, and if they only read newspapers that similarly represent that same world view, and if their social circles are similarly limited to the like-minded, they are insulating themselves from alternative ideas that might challenge their mindset and perhaps lead them to modify it. They can easily come to regard their world-view as the only reasonable one, one shared by all decent people.

This rather resembles the way in which social classes use shared behavioural practices to exclude outsiders.

*I thought everyone ate asparagus with the fingers.*
*Nobody puts coloured lights on a Christmas tree.*

(Only people who share our values and habits are worth considering. Those who do are 'everyone', and someone who does not is
'nobody'.)

What is mildly amusing in the world of class manners becomes
counter-productive in the world of ideas and argument. Those who
insulate themselves from alternatives commit the fallacy of supposing
that the ideas they share with their like-minded colleagues and friends
are the truth, rather than just contenders for that title that have to be
tested against their opposition.

As for your use of the fallacy, when people suggest ideas other than
the ones universally agreed upon by you, your friends, your gym class
and the other customers at your wholefood store, learn to shrug your
shoulders and roll your eyes at the notion that such outlandish notions
should even make it into the discussion.

# Irrelevant humour

The fallacy of irrelevant humour is committed when jocular material
irrelevant to the subject under review is introduced in order to divert
attention away from the argument.

*My opponent's position reminds me of a story . . .*

(Which will not remind the audience of the argument.)

While humour entertains and enlivens discussion, it also distracts. The
fallacy does not lie in the use of humour but in its employment to
direct attention away from the rights and wrongs of the matter in hand.
A joke might win an audience, but it does not win an argument.

*A member of parliament, Thomas Massey-Massey, was introducing
a motion to change the name of Christmas to Christ-tide, on
the grounds that mass is a Catholic festival, inappropriate to a*

*Protestant country. He was interrupted by a member opposite who asked him how he would like to be called 'Thotide Tidey-Tidey'. The bill was forgotten in the uproar.*

The hustings heckler is the great exponent of this fallacy. His warblings accompany parliamentary election meetings, often drowning out any reasoned argument for the very good reason that they are a lot more interesting and quite often of a higher intellectual level. A few of them achieve the immortality of the book of quotations as 'anonymous heckler', especially if their interjection has prompted an even better reply from the candidate. Lloyd George, Winston Churchill and Harold Wilson all showed adroitness at turning a diversionary joke back upon its user.

QUESTIONER: *What do you know about agriculture? How many toes has a pig?*
NANCY ASTOR: *Why don't you take off your shoes and count them?*

Often cited as a classic of irrelevant humour is the joke by Bishop Wilberforce when debating evolution against Thomas Huxley. Pouring scorn on evolution, the bishop asked Huxley:

*You claim descent from a monkey; was it on your grandfather's or grandmother's side?*

(Huxley's reply is also considered to be a classic put-down. He saw no shame in being descended from a monkey, but described the man he *would* be ashamed to have as an ancestor; a man who despite his learning sought to obscure by means of aimless rhetoric and appeals to prejudice . . .)

The problem for the user of rational argument is that a guffaw is as difficult to refute as a sneer. The audience enjoys the entertainment more than the argument. A speaker for a religious sect would regularly invite his audience to supply any biblical quotation which conflicted with his view of things. When members of the audience obliged, as they often did, he would always reply:

*That sounds more like Guinesses talking than Genesis.*

(The volunteer was invariably discomfited by the gale of laughter.)

Those who set out upon the trail of public debate should carry a knapsack full of custom-built jokes ready to toss before an audience in times of need. At the very least, the wave of mirth washing over your victim slightly lowers his authority, even while it gives you time to think.

The ability to produce irrelevant humour on the spur of the moment is a product of wit and experience. Many years spent in the debating chamber of a university will sharpen your ability to think on your feet. The joke need not even be a clever one if delivered with style. I once saw a speaker making a perfectly valid point about sales to authoritarian states of airplanes which could carry nuclear weapons. He was floored by an interjection which suggested that wheelbarrows could do the same.

An undergraduate in the process of being censured for high crimes and misdemeanours took all of the force out of the attack by facing his audience solemnly and saying:

*I wish to accept censure, and to couple with it the name of my mother, who also thinks I've been a very naughty boy.*

(Collapse, amid uproar, of prosecution case.)

# It's worth it if it saves lives

This line can be used to justify almost anything, no matter what the cost, and therein lies its error. In its more extreme form people say, 'It's worth it if it saves just one life.' Sometimes people say, 'You cannot put a price on human life,' and the sentiments are more admirable than the sense. The point is that we are not prepared to spend the entire resources of the nation to save just one life. Having made that insight, we are then talking not about principle but about price. We are asking how much it is worth spending in order to save a life.

We could decide to close down some schools and universities, and spend the money instead on road safety and health services so that lives were saved, but few people would think that a worthwhile trade-off, considering what schools and universities achieve. When we decide how much to spend in order to save a life we have to think what that money could otherwise buy, and whether the sacrifice is justifiable.

A life might have near infinite value to a person's partner or parent, but would not have that value to people who did not know the person concerned. Many people might eat plain foods for a year in order to spend the money thus saved on saving someone they loved, but not many would be ready to do it in order for the money to be used to save unknown lives by making our rivers and lakes safer for swimmers.

Railways could charge a great deal more for their tickets and spend it making them safer than they are. The train and track companies actually use a rough calculation that it is worth spending just under a million pounds to save a life. If the improvements would cost £10m and save only six lives, they would not be considered worth doing. In the past, when sometimes passengers were killed falling from trains, the fitting of secondary locks was considered and calculations showed they could expect to save lives at a cost of £50,000 each. This was thought to be a good trade-off and was implemented. Few people, however, would be happy to see train fares trebled to fund safety improvements that saved only a tiny number of lives.

The problem is most acute, and the fallacy used most frequently, in Britain's health service. People want many more resources put into saving their loved ones through extraordinary and expensive procedures. But the NHS personnel understand that money spent on one patient cannot also be spent on another. The resources to keep alive a premature newborn baby cannot also be spent on a liver transplant for a young man or for cancer chemotherapy for a middle-aged woman. The cost of saving a life can be set in terms of the other lives that will be lost by doing it.

Despite the sense of such calculations, the emotional appeal of the fallacy is so strong that people are thought hard-hearted and cold if they list the trade-offs dispassionately. This means that in practice you can win many arguments by claiming you will be saving lives. It is quite possible that someone somewhere has proposed the erection of fencing around the entire UK coastline at a cost measured in trillions.

> 'After all, if it saved just one person from falling into the sea and drowning it would be worth it.'

(No it wouldn't.)


# Lapidem, argumentum ad

Bishop Berkeley expressed the view that matter does not exist separately from the perception of it. When Boswell told Dr Johnson that this was an idea impossible to refute, the good doctor's response was to kick against a stone so that his foot rebounded. 'I refute it thus', he said. He was not so much refuting it as ignoring it, because the evidence for the existence of the stone, including the sight, sound and feel of a kick against it, is all perceived by the senses.

Dr Johnson's treatment has given us the name of the *argumentum ad lapidem,* the appeal to the stone. It consists of ignoring the argument altogether, refusing to discuss its central claim.

> *He's a friend of mine. I won't hear a word spoken against him.*

(Top marks for loyalty; none for knowledge.)

An argument or piece of evidence cannot be dismissed because it fails to conform to an existing opinion. Much as we might like to toss out material which offends our ordered view of things, it is a fallacy to suppose that we can do so without cost. By refusing to admit material which may be relevant to a sound conclusion, we proceed in ignorance. Ignorance is more reliable as a source of bliss than of correctness.

The *argumentum ad lapidem* is most appropriately named after Dr Johnson's use of it, for it was one of his favourites. His reasoned and balanced view on the freedom of the will, for example, came out as:

*We* know *our will is free, and there's an end on't.*

(It does tend to finish an argument, as it is meant to.)

Jeremy Bentham described all talk of natural rights as nonsense, and talk of inalienable natural rights as 'nonsense on stilts'. So much for the American Declaration of Independence.

There are always plenty of stones to kick in fields where proof has no footing. Wherever a belief is indemonstrable, its adherents can use the *ad lapidem.*

*Reason is no guide; you must open your heart . . . and you will* know.

(This is not terribly useful to outsiders looking for the truth of a thing, comforting though it may be to those who *know.*)

This fallacy occurs in a university setting far more than one might suppose. It is often argued quite seriously in quasi-academic circles that certain books should not be allowed on campus because they propagate error. Speakers are quite frequently shouted down because their audience *know* they are speaking falsehoods, and do not need to hear the argument. Some student unions actually make a policy of the *ad lapidem,* refusing to allow a platform on campus for known error – a category which can even include members of the elected government of the country.

A charming version of the fallacy emerged from the pen of Herbert Marcuse. Now forgotten, although he was a high priest of student radicals in the 1960s, his *Critique of Pure Tolerance* made the interesting point that tolerance can be repressive because it permits the propagation of error. How could we recognize error in order to stop it? Easy. Guess who was going to tell us.

When you yourself employ the *ad lapidem,* you must do so with a total assurance which suggests that the person who raised the offending fact or argument is totally beyond the pale. Like the judge who once convicted a jury for 'going against plain evidence', you should make it clear that he is going against all reason. Your opponent, by going beyond every received opinion and every canon of decency, has rendered his opinion totally unworthy of discussion.

*Liberty of expression is one thing; but this is licence!*

('Licence' means liberty you don't approve of.)

Where you have control of events, you can afford to be less subtle: 'I don't care what time it is. Get to bed this minute.'

# Lazarum, argumentum ad

The poor may indeed be blessed; but they are not necessarily right. It is a fallacy to suppose that because someone is poor that they must be sounder or more virtuous than one who is rich. The *argumentum ad Lazarum,* after the poor man, Lazarus, takes it that the poverty of the arguer enhances the case he or she is making.

*The guru has nothing to gain by lying or fooling anyone; all he has are the nuts that he lives on.*

(And the ones that he teaches.)

Poverty does not contribute to the soundness of an argument, any more than riches do. The fallacy consists of giving attention to the person instead of to the contentions which he or she is putting forward. It may well be that the poor are less exposed to the temptations of affluence, but it may equally be that the rich are less distracted by disease, hunger and degrading toil, and the temptations to escape them. Even if we take it that a person who eschews wealth is not acting for material gain, we should remember that there are other ways of achieving satisfaction. 'All power is delightful', we are told, 'and absolute power is absolutely delightful.'

Although we should not take account of the circumstances of the arguer, the *ad Lazarum* is deeply engrained into our thinking. We tend to suppose that the poor have less opportunities for error, having less opportunity, full stop. The literature of our culture compensates them for their poverty with extra measures of wisdom and virtue, sometimes of beauty.

*With her clogs and shawl she stood out from the others.*

(It could just have been malnutrition, though.)

The poor are probably more likely to prefer the means of acquiring real education, health and respite from an arduous life than to want the phantoms wished on them in the rose-tinted imaginations of detached observers.

The politician who astutely recognizes that most of his constituents are poor will often go to extraordinary lengths to feign a similar poverty, thereby hoping to command respect. His limousine is left at the frontier with his well-cut suit, as he changes down to the car and clothes of his constituents. Those self-same electors, did he but know it, probably regard him as no better than themselves, and reserve their admiration for the guy with the flash car and the swanky outfit. The point is that the *argumentum ad Lazarum* is a fallacy which appeals to the well-to-do. The real poor have no time for it.

*The best view I ever heard on this was told to me by a simple, honest woodcutter . . .*

(Who was probably smart enough not to depend on the views of woodcutters . . .)

Woodcutters, like aged peasants with weatherbeaten faces, should be lined up in orderly squadrons in support of your arguments. A few simple fishermen should act as outriders, with a score or two of wise old washerwomen in reserve. Their faces, lined by experience, should nonetheless reflect an inner placidity and acceptance of life. The views which you put forward were, of course, gained from sources such as these.

*He puffed reflectively on his pipe, then looked at me with those strangely quiet eyes. He told me that, although poor himself and honest, he had always reckoned that deficit spending by government could stimulate production by priming demand, and similarly . . .*

(If he's so sincere, how can he be wrong?)

# Loaded words

It is possible to influence the outcome of a judgement by the deliberate use of prejudiced terms. When the words used are calculated to conjure up an attitude more favourable or more hostile than the unadorned facts would elicit, the fallacy used is that of loaded words.

*HITLER SUMMONS WAR LORDS!*

*M. DALADIER CONSULTS DEFENCE CHIEFS*

(The two headlines tell us the same thing: that the leaders of Germany and France had seen the heads of their armed forces. In Germany these are 'war lords', but in France they are 'defence chiefs'. The German leader is simply 'Hitler', without title, and he summons his men imperiously. Daladier, however, is a monsieur, and being a good democrat, 'consults'.)

Near synonyms carry subtle nuances of meaning which can be used to influence attitudes to the statement which bears them. The fallacy derives from the fact that these attitudes are not part of the argument. They were conjured up illicitly to achieve more effect than could the argument alone. The extra nuances and the response to them are both strictly irrelevant to establishing the truth or falsehood of what is being said. Language abounds with ways of putting our own attitudes into a description in order to elicit a response from others. People may be forgetful or negligent; they may be steadfast or unyielding; they may be confident or arrogant. Many of these terms are subjective: they depend for their accuracy on the feelings of the observer and on how he or she interprets the situation. A fair argument requires a conscious effort to put forward the case in terms which are reasonably neutral.

*Once again Britain has been found sucking up to dictatorships.*

(Or maintaining friendly relations with strong governments. Note how 'found' implies that they were discovered in a guilty secret.)

The judge's bench, as he directs the jury, is good territory for loaded words to roam on. English law, through a tiresome oversight, gives the jury the right to decide the verdict. Many a judge will help to fill this gap in legal procedure by choosing words to help the unfortunates in their deliberations.

*Are we to believe the word of this snivelling, self-confessed pervert, or that of a man whose reputation is a byword for honour and integrity?*

(If you had thought of doing so, this is a good point at which to change your mind.)

There is a series of verb conjugations which brings out the different loading a speaker will apply to words describing him, the person he speaks to, or an absent third party. Thus: 'I am firm; you are stubborn; he is a pig-headed fool.'

Descriptions of contests can invite us to take sides by the choice of terms, rather than by the events they report.

> *Scotland stole a goal in the first half, but England's efforts were well rewarded in the second half when . . .*

(Guess which side of the border the reporter comes from?)

What goes for the sports section applies even more to the leader page.

> *The public can distinguish Labour bribes from Tory pledges.*

(They can certainly distinguish whose side the writer is on.)

Public affairs programmes on television are great fun for the connoisseur of loaded words. There is an unfortunate conflict of interests. They want to present material to make you share their prejudices; their authority requires at least some semblance of objectivity and balance. While blatant bias does occur, the satisfaction comes in spotting the loaded words at a slightly more insidious level. Which side has 'terrorists', for example, and which has 'freedom-fighters?' Which countries have a government and which a regime?

When you are in the situation of trying to *persuade* people, you will find loaded words most useful. Your verbal picture shows the bleak outlook of one alternative, and contrasts it with the rosy setting which results from the other. Your listeners need never know that you could have done it just as easily the other way round.

> *Would you rather believe the careful words of an internationally respected columnist, or the incoherent ramblings of a well-known hack?*

*Are you not moved by the just case which is even now being voiced by thousands of concerned demonstrators outside this very building?*

*I'm not going to be taken in by the bleatings of a mob.*

When describing actions, remember to load your words in such a way that even to observers who know nothing of the facts, there will be an obvious distinction between your prudent investments and the reckless spending of others, between the modest perquisites to which you were entitled and the wholesale embezzlement in which they have engaged. Your dispassionate testimony should contrast well with their frenzied diatribe.

# Misericordiam, argumentum ad

While pity is an admirable human quality, it does not provide the best basis for argument. When we turn to pity instead of reasoned discourse to support a particular contention, we commit the *argumentum ad misericordiam.*

*In asking yourself if this man is to be convicted, ask yourself what it will mean for him to be locked up in prison, deprived of his liberty, and turned into an outcast from humanity.*

(The question is whether he is guilty or not, not what conviction will do to him.)

When we are called upon to settle questions of fact, we should be weighing up the evidence on each side and attempting to arrive at the truth. The introduction of pity does nothing for the argument. While it might reasonably influence our actions, it should not influence our judgement. The consequences to various parties of the truth or falsehood of a statement does not bear on that truth or falsehood.

Whether a man is sent to prison or to the South Seas for a holiday does not alter the fact itself. An *ad misericordiam* is committed if pity is appealed to in the settlement of questions of truth and falsehood.

> *Can we continue to afford Jeeves as our groundsman? Look what will happen if we don't. Imagine the state of his wife and his children with Christmas coming up and the cold snows of winter about to descend. I ask instead, 'can we afford not to employ Jeeves?'*

> (Yes we can. Of course, we might *decide* to afford him, which is quite a different thing.)

Quite apart from its use in courts of law – where no self-respecting defence lawyer will venture without his handkerchief – the *ad misericordiam* pokes its head into any argument where facts have consequences. No one would allow the possible fate of an individual to influence our conviction about so obvious a fact as $2 + 2 = 4$, but where there is less certainty we might be tempted to allow our pity to give the benefit of the doubt.

Hearts and flowers are a prerequisite of public policy. No question of simple fact can be settled without consideration of the effect it might have on the sick, the old, the feeble, the blind and the lame.

> *If we decide that foreign aid is ineffective, and does not raise living standards, then we are condemning people in the poorer countries to a life of degrading poverty, squalor and disease.*

> (If foreign aid is ineffective, the fact condemns them to these consequences. Maybe we should do something else about it.)

The appeal of the *ad misericordiam* is in our recognition that pity should have a place in guiding our actions. The point of the fallacy is that it has no place in our determination of truth and falsehood. When it steps from one territory to the other, reason changes place with it.

Its allure is hard to resist. The whole of Dickens' *A Christmas Carol* is one giant *argumentum ad misericordiam.* Here is Scrooge, making an honest living, assailed (along with the reader) by the appeal to pity. Bob Cratchit commands a skill as a clerk and scribe, and is perfectly free to seek employment elsewhere at market rates if he is dissatisfied with what Scrooge offers. But no; ghosts rise up to torment his employer with the *ad misericordiam,* and the hapless Scrooge is morally compelled to reach a decision quite contrary to economic reality. A more valid response to this treatment would have been 'Bah! Humbug!'

You will have a great time making your opponents squirm under the *ad misericordiam.* Your audience is not too interested in the fine distinction between fact and fiction, so you can easily make those who reach different conclusions about the truth of things seem like the most hard-hearted of Victorian landlords for doing so.

*If you really believe that high wages keep teenagers from getting jobs, then all I can say is that you will have on your conscience the thousands of poor families who struggle to find the means for life's necessities. May God have mercy on your soul!*

(Even if he does, the audience won't. When faced with this treatment, turn it right back. What about the suffering and humiliation of those poor teenagers, unable to find work because of your heartless opponent? You can't expect to win with duelling pistols when your opponent is using a howitzer.)

# *Nauseam, argumentum ad*

Simple repetition of a point of view does nothing by way of supplying additional evidence or support. Yet it can erode the critical faculty. There is a completely mistaken supposition that a thing is more likely to be true if it is often heard. The *argumentum ad nauseam* uses constant repetition, often in the face of massive evidence against a contention, to make it more likely to be accepted.

> *Just the place for a Snark! I have said it thrice; what I tell you three times is true.*

> (Whereas in fact if someone repeats the same thing three times it is because he has nothing else to say.)

The point is that repetition adds nothing at all to the logic. It is done in an attempt to persuade an audience, either by wearing down their resistance, or by deceiving them into supposing that objections have somehow been dealt with. Since they add nothing, the extra versions are irrelevant to the consideration, and fallaciously appeal to psychological factors rather than to reason.

> *'Please sir, it wasn't me!'*
> *'But this is your catapult, Smith Minor.'*
> *'Please sir, it wasn't me!'*
> *'And witnesses saw you pick up the stone.'*
> *'Please sir, it wasn't me!'*

> (This could go on indefinitely, unless the heavy hand of an *ad baculum* cuts it short. We can all spot that Smith Minor would have done his case more good if he had been able to find anything else to say. Would we spot it if he simply kept saying 'Socialism means rule by the workers', however?)

Utterly discredited political credos, which adherents cling to for other than intellectual reasons, are supported by the *ad nauseam* fallacy. If an economic system brings general prosperity and gives ordinary people access to the things which were once the prerogative of the rich, it is quite difficult to make out a case that this is exploitation. Fortunately, one does not have to. The *ad nauseam* effect means that the charge can simply be repeated over and over again without argument or evidence. Eventually, some people will fall for it.

Advertisers have long been life members of the *ad nauseam* society. They know that a specious claim acquires credibility and force if it is repeated often enough. They know the importance of building up not a rational conviction but a habit of association.

> *It washes whiter than bleach; that's whiter than bleach; yes, whiter than bleach.*

(What they tell us three times is true.)

Many of the proverbs we hear in childhood are dinned into us so many times that we often come to suppose that there must be truth in them. This assumption seems able to survive all of the contrary evidence which life thrusts before us, and in some cases survives a simultaneous belief in contradictory proverbs. It is quite hard to look before you leap without being lost through hesitation, and while many hands make light work, they do tend to spoil the broth. All of which shows the power of the simple *ad nauseam.*

To use the *argumentum ad nauseam* is easy enough: all you have to do is to repeat yourself. It is harder to recognize the situations where it might succeed. The general rule is that constant repetition over a long period of time is more effective than short bursts. You must be totally impervious to arguments against you, always reiterating the same point. This not only bores your audience to tears, it also instils in them the futility of opposing you. And when they give up in total weariness, observers will begin to suppose that they can no longer counter your claims.

The civil servant advising his minister provides a case-study of the *argumentum ad nauseam*:

> *But Minister, as I have been explaining for two years, there is no way in which we can cut the administrative costs of this department.*

*Every single job is vital to our efficiency. The man we hire to scour the building picking up used paper-clips, for example . . .*

(And the *ad nauseam* cracks the minister before the minister cracks the problem.)

If you aspire to expert rank, however, study closely the form exhibited by the minister himself at the dispatch-box:

*I responded to charges of ministerial dereliction of duty on 9 November by saying that I had nothing to add to my statement of 4 June. I would not care to expand on that at this time.*

(Please sir, it wasn't me!)

# Non-anticipation

The fallacy of non-anticipation consists of supposing that everything worth doing or saying has already been done or said. Any new idea is rejected on the grounds that if it were any good, it would already be part of current wisdom. Proposals are rejected because they have not been anticipated.

*If tobacco really is so harmful, how come people didn't ban it years ago?*

(They didn't know. Nowadays more people live long enough to experience the adverse effects, and we now have more techniques for measuring such things.)

The central assumption of the fallacy is unwarranted. Progress is made on several fronts, including the scientific and the social. New ideas are constantly being adopted, and there is no justification for supposing that our ancestors would have found them all. The presumption that they did intrudes irrelevant material into the argument.

Wise though the sages of old probably were, we can no more presume in them the totality of wisdom than we can assume complete stupidity.

*If breakfast television is all that good, why has it taken so long for it to appear?*

(Because we didn't realize that people wanted even more pap with their morning milk.)

It is not just products and processes which are revolutionized by invention; the same is true of changes in our living patterns.

*People didn't need these long Christmas holidays years ago, why should they now?*

(They probably did need them years ago; they just couldn't afford them. The same fallacy would have supported, and no doubt did, the continuance of child labour in mines and factories.)

The fallacy of non-anticipation is a great comfort to those who, while possessing a conservative disposition, cannot actually think of any arguments against the changes which are put forward.

*Mr Chairman, this proposal has been kicked around for more than twenty years. If there were any merit in the idea at all, it would have been implemented long before now.*

(The beauty of this is that your current rejection will serve as extra 'evidence' against it in the future. Perhaps the reasons for past rejection were equally frivolous.)

To give added effect to the fallacy, you can enumerate some of the phantom legions who could have taken up the idea but did not. Their numbers appear to be ranged against the idea, like yourself, even though they might simply never have thought of it.

*Are we to assume that we are cleverer than the thousands of the very learned and competent people over the years who could have acted on a proposal such as this, but wisely refrained from doing so?*

(Any more than Beethoven was cleverer than the millions who could have written his symphonies but did not do so?)

You will find the fallacy extraordinarily useful in resisting trends towards emancipation. After all, if there were any merit in having women and children participate in decisions, would it not have been discovered long ago? The same approach will help you to stand up against independent holidays, dining out, taking exercise or eating courgettes.

*If there were any connection between drinking eight pints of beer a day and obesity, don't you think that countless beer drinkers would have seen it before now?*

(Why should they? They cannot even see their own toes.)

## *Novitam, argumentum ad*

If it is a fallacy to suppose that age is a guide to correctness, it is also fallacious to suppose something to be more right simply because it is new. The *argumentum ad novitam* makes the mistake of thinking that the newness of something is a factor contributing to its soundness. To hear support urged for something because it is new is to hear the *ad novitam* being used.

*These new tower-blocks are the coming thing. We should build some ourselves.*

(Their newness did not stop them brutalizing the landscape of cities or the lives of their tenants.)

Some people are surprised to find that both newness and oldness can be used fallaciously in support of a contention. In fact they appeal to contradictory traits in all of us. We like the security of the traditional, and we like to be fashionable and up-to-date. Either of these can be used as fallacies if we try to make them support claims which should stand or fall on their merits. The *ad novitam,* like its *antiquitam* counterpart, introduces the irrelevant fact of the age of the proposition as a means of influencing its acceptance. Because the newness does not, in fact, contribute to its rightness, a fallacy is committed by appealing to it.

There was a time when the *ad novitam* found as welcome a home with progressive reformers as its sibling, *ad antiquitam,* did with conservatives. Those were the days of constructing a brave new world. Times change, however, and the *ad novitam* now builds its nest amongst conservatives. It settles down comfortably amid calls for the rejection of 'the old ways which have failed' and for 'looking truly fit for the twenty-first century'. Meanwhile the *argumentum ad antiquitam* stirs uneasily as it sees progressives looking back to the good old days of social reform.

Advertisers have used the word 'new' as a reflex appeal to the *ad novitam* for many years. Assuming that the public equated new products with new progress, everything from washing-powder to toothpaste has been 'new, improved'. Breakfast cereals were forever new, with the main innovation being the increasing resemblance of the contents to the cardboard of the packet. Great were the shockwaves in the advertising world when cereals started to appear which were positively old in style. In faded brown packets, they promised old-fashioned goodness, and rapidly gained sales. The bold attack of the *ad antiquitam* sent the *ad novitams* back on the ropes. All kinds of products came out with old-fashioned presentation. 'Just as it used to be' was the slogan, with sleepy scenes and pictures of cobwebs on the packets. In Britain, Hovis bread, instead of being new and improved, featured sepia-tinted ads of rural simplicity.

Both fallacies have powerful appeal, but *ad novitam* had gone too far. Now there is a balance between the two. The simplest country boy wears clothes looking roughly like a space-suit, while those brought up in Glasgow tenements now look back on entirely false childhood memories of country smells and fresh brown eggs.

When using the *ad novitam,* remember the conflicting appeal of the two fallacies, and confine it to areas where the *ad antiquitam* is unwelcome. You cannot support housing because it is new, since people will prefer the old. But you can support economic theories because they are new. After all, what good ever came of the old ones?

Just as yours is the 'new economics', so are your social and moral convictions part of the 'new awareness'. An audience would much prefer to be brought up-to-date and given new information, rather than being hectored to change their minds.

> *Are we to continue in the ways of the old acquisitiveness by allowing commercial development on the site, or are we to respond to a new awareness of social needs by building a modern community centre for the unemployed?*

(With arguments like this, you'll win easily. You'll get a community centre for those who would have been employed by the commercial development.)

## *Numeram, argumentum ad*

Not many people like to be out on a limb. Many prefer instead the comfort of solid numbers behind them, feeling there is less possibility that so many others could be mistaken. The *argumentum ad numeram* wrongly equates the numbers in support of a contention with the correctness of it. Ideas which have mass support are not necessarily more likely to be right; but the *ad numeram* supposes that they are.

> *Fifty million Frenchmen can't be wrong!*

(A glance at the history of that nation will show that they very often have been.)

The fallacy lies in the fact that the rightness or wrongness of a contention is neither helped nor hindered by the numbers in support. Many people are quite often wrong about even simple things, and received wisdom is not to be equated with factual knowledge. Simple observation, such as that which shows planets and stars turning across the skies, can be an unreliable guide, no matter how many millions attest to its truth.

*Everybody's smoking Whifters, why don't you?*

(Because he thinks everybody is stupid.)

The *ad numeram* can appeal to general numbers, or more insidiously, to the numbers of those you respect. You might be more impressed by the proportion of top people taking *The Times,* than by the numbers backing Britain's biggest daily sale. The question to ask yourself is whether the numbers add anything to the claim.

*We have an argument here about whether Ballasteros ever captained a European golf team. Let's settle it democratically.*

(And before you jump off a building, make sure you have enough votes to carry repeal of the law of gravity.)

If ideas were decided by numbers, no new ones would ever be admitted. Every new idea starts out as a minority viewpoint and gains acceptance only if the evidence for it wins converts over from the prevailing view. If numbers are the test, then Giordano Bruno was wrong when he said the earth moved around the sun, and the authorities were right to burn him at the stake.

*We have to give him a fair trial before we string him up. All those who say he did it shout 'aye!'*

(Amazing proof! Sounds kinda unanimous to me.)

The *ad numeram* provides an excellent defence of established attitudes.

> *If it's not true, then why have so many millions of people believed in it for so many centuries?*

(Easy. We all make mistakes.)

The *ad numeram* is the special fallacy of the demagogue and the mob orator. Those who govern us tend to form a special class whose outlook and assumptions are not commonly shared. They often come from a milieu in which the pressures of poverty, overcrowding and crime bear rather less upon them than they do upon most. This gives the demagogue the opportunity to appeal to numbers in support of ideas which find little echo in government. On subjects such as capital punishment or race relations, he can appeal to the agreement of large numbers on his side as evidence of a conspiracy of silence by the governing elite.

> *Every opinion poll shows that public whipping is the best remedy for those who commit crimes of violence.*

(And if you asked them, they'd probably say the same about garotting and disembowelment. They could be just as wrong or, indeed, right.)

The *ad numeram* is a fallacy to be used with passion. In its ideal setting you would be haranguing a rabble of 600 people armed with blazing torches outside a corn merchant's house during a famine. Even in print, you should not turn an *ad numeram* into a clinical counting of heads, but conjure up outrage that the obviously correct view of so many should be ignored.

When your side is in an unfortunate minority, the technique is to quote from the past, when your lot were on top, or from foreign countries where you do have a majority to back you. Sweden is an excellent source for majorities in favour of the most bizarre things.

*Are we to say that all Swedes are fools? That the people of the world's most enlightened country don't know what they are talking about?*

(Yes.)

# Omitting the benefit

Cost-benefit analysis is a perfectly legitimate activity. When an activity is examined it involves counting up the benefits it yields and setting them against the costs it entails. Neither the benefits nor the costs need to be monetary. They can be of any advantages and disadvantages resulting from the activity. Normally this cost-benefit analysis is done to establish whether an activity is worthwhile, meaning that the gains from it outweigh any adverse consequences. It can be done for current activities or for proposals for future activities. If a course of action is suggested, it makes sense for the likely consequences, both good and bad, to be examined and set against each other.

A surprisingly common fallacy consists of omitting the benefits produced by an activity or a proposal, and counting only the adverse consequences. If only the negative aspects are counted, people are unlikely to form a realistic assessment of how worthwhile it might be. Most activities involve some costs, perhaps even trivial ones such as the time spent doing it that might have been employed doing something else. To list only the costs is to miss out on the more positive consequences deriving from it.

*We should ban mountaineering because it is too dangerous. Several people are killed or injured doing it every year.*

Yes, this is true, but what about the benefits? What about the thrills and the excitement that the danger generates? What about the feeling of achievement when the climber reaches the summit and looks down?

They might or might not compensate for its downside, but they should be taken into consideration.

Various campaigns against things such as drinking, smoking, eating sugary foods or burgers often fail to make an impact because they do not take account of the reasons people have for doing those things. Campaigns might be more successful if they recognized that some people like doing those things and attempted to address that fact.

*We should help people to live longer by putting heavy taxes on sugary drinks, salted foods and fatty burgers as well as on alcohol and tobacco.*

(People impoverished after paying all those taxes may find it doesn't help their longevity.)

Using the fallacy yourself is easy. You simply pour scorn on whatever anyone proposes by listing all of the horrendous things that could happen if it were to be done, without mentioning any of the gains that would result from it.

*Put life-jackets on ships? Think of the cost and the other things you could have done with that money! How bulky and awkward they would be to wear; they might lead people to fall over and injure themselves. One of them might save the life of a future murderer; would you want that guilt on your hands?*

And so on . . . Just don't mention the hundreds of lives they could save.

# One-sided assessment

Many of the decisions we are called upon to weigh up have both advantages and drawbacks. The fallacy of one-sided assessment is fallen into when only one side of the case is taken into consideration. Decisions usually require both pros and cons to be taken account of, and a preference made for the side that wins on balance. To look at one side only is to avoid judgement of that balance:

*I'm not going to get married. There would be all that extra responsibility, not to mention the loss of my freedom. Think of the costs of raising children and putting them through college. Then there are the increased insurance premiums . . .*

(If that really were all, no one would ever do it.)

It is equally possible to look only at the positive side.

*This encyclopedia is one you will be proud to own. Your friends will admire it. Your children will benefit. You will learn from it. It will complement your bookshelf!*

(On the other hand, it will cost you a LOT of money.)

Either way, the fallacy of one-sided assessment is committed. By looking only at the objections, or only at the advantages, we are excluding material which bears on the decision, and which should be taken account of. The omission of this relevant material from the argument is what is fallacious about one-sided assessment.

One-sided assessment is not a fallacy when space is allocated for the other side to be presented similarly. There is in Anglo-American culture an adversarial tradition, which has it that if each side has the strongest case put forward, then a dispassionate observer is likely to arrive at a fair judgement. We therefore expect a counsel to put only the evidence for acquittal, and a trade union negotiator to put only the case for an increase, because we know that there will be someone else putting the other side. It would be one-sided assessment if those making the *judgement* considered one side only.

*Let's not go to Ibiza. Think of the heat, the mosquitoes and the crowds.*

(On the other hand, what about the lovely sunshine, the cheap wine, the excellent food and the low prices?)

Life's judgements often call for trade-offs. Those who have made their balance and come down in favour are apt to try to persuade others by emphasizing only the positive side. The unwary should remember that their own scale of values might call for a different judgement, once they consider all of the factors.

*All of the arguments support the new road. It means progress; it means prosperity; it means a future for our town!*

(And it really is rather unfortunate that they have to pull your house down to build it.)

There is a clever version of one-sided assessment which you should use when persuading others to agree with your judgement. This involves making a purely token concession to the case against you, by referring to one of the weaker arguments on the other side before you launch into the overwhelming arguments in favour. This polishes your case by adding to it the gloss of apparent objectivity.

*Of course, if we bought a bigger car, we'd need to make new seat covers. But think of the convenience! All the shopping would go in the back; we could use it for holidays; you could pick up the children in comfort; and its extra speed would cut down our journey times.*

(Sold, to the gentleman with the fallacy.)

# *Petitio principii*

The fallacy of *petitio principii*, otherwise known as 'begging the question', occurs whenever use is made in the argument of something which the conclusion seeks to establish. The *petitio* is a master of disguise and is capable of assuming many strange forms. One of its commonest appearances has it using a reworded conclusion as an argument to support that conclusion.

*Justice requires higher wages because it is right that people should earn more.*

(Which amounts to saying that justice requires higher wages because justice requires higher wages.)

It might seem to the novice that *petitio* is not a fallacy to take for a long walk; it seems too frail to go any distance. Yet a short look at the world of political discourse reveals *petitios* in profusion, some still running strongly after several hundred years. It is quite difficult to advance arguments for a commitment which is in essence emotional. This is why politicians deceive themselves accidentally, and others deliberately, with a plethora of *petitios.* The political petitio usually appears as a general assumption put forward to support a particular case, when the particular case is no more than a part of that same assumption.

*The British government should prohibit the sale of the Constable painting to an American museum because it should prevent the export of all works of art.*

(It looks like an argument, but the same reason could be advanced for each particular work of art. Adding them up would tell us no more than that the government should prevent the export of all works of art because it should prevent the export of all works of art.)

Argument is supposed to appeal to things which are known or accepted, in order that things which are not yet known or accepted may become so. The fallacy of the *petitio principii* lies in its dependence on the unestablished conclusion. Its conclusion is used, albeit often in a disguised form, in the premises which support it.

All arguments which purport to prove the unprovable should be carefully scrutinized for hidden *petitios*. Arguments in support of ideologies, religions or moral values all have it in common that they attempt to convince sceptics. They also have it in common that *petitios* proliferate in the proofs.

*Everything can be defined in terms of its purpose.*

(Never be surprised when a discussion starting like this ends up 'proving' the existence of a purposive being. If things are admitted at the outset to have a purpose, then a being whose purpose that is has already been admitted. This is a *petitio principii* disguised as a proof.)

When using the *petitio* yourself, you should take great care to conceal the assumption of the conclusion by skilful choice of words. Particularly useful are the words which already have a hidden assumption built into them. Words such as 'purpose' fall into this group. Philosophers always go into battle with a huge stockpile of these words, especially when they try to tell us how to behave. The obligations they wish to impose upon us are hidden away in words like 'promise'. It looks like a straight, factual thing, but it has an 'ought' tucked away in its meaning.

The important thing to remember about the *petitio* is that it is supposed to look like an argument in support of a case. You should therefore spatter it with argument link words such as 'because' and 'therefore', even if it is no more than a simple rewording.

When pushed into a corner you can often effect a dramatic escape with a well-chosen *petitio* by combining *both* the assumption of a general truth, *and* a rewording of the conclusion.

*We should not sell arms to Malaysia because it would be wrong for us to equip other nations with the means of taking human life.*

(This looks and sounds like an argument, but it is really just a clever way of saying that we should not sell arms to Malaysia because we should not sell arms to anyone.)

# Poisoning the well

The most attractive feature of poisoning the well is that the opposition is discredited before they have uttered a single word. At its crudest, the fallacy consists in making unpleasant remarks about anyone who might disagree with a chosen position. When some willing victim steps forward to dispute that position, he only shows that the unpleasant remarks apply to him,

*Everyone except an idiot knows that not enough money is spent on education.*

(When someone comes forward to suggest that enough money *is* being spent he identifies himself to the audience as the idiot in question.)

The whole discussion is fallacious because it invites acceptance or rejection of the proposition on the basis of evidence which has nothing to do with it. The claim is only an insult, offered without evidence, and does not have to be accepted. Even if it were true, we would still have to examine the argument on its merits.

Closer inspection shows that poisoning the well is a highly specialized version of the *ad hominem* abusive. Instead of insulting the arguer in the hope that the audience will be led to reject his argument the well-poisoner sets up the insult for anyone who might argue. It is cleverer than simple abuse because it invites the victim to insult himself by drinking from the poisoned well. In doing so, it discourages opposition.

*Of course, there may be those with defective judgement who prefer buses to trains.*

(There may be those who take into account such factors as price, cleanliness, convenience, and running on time. To admit the preference now, however, would be owning up to defective judgement.)

In its crude and simple form, poisoning the well is seen to be great fun and can engender spectacular coups of withering scorn. A version which is only slightly more subtle appears in a game called 'sociology of knowledge'. To play the game, one player starts by asserting that everyone else's view about society and politics is only the unconscious expression of their class interest. Next, he shows that for specialized reasons this analysis does not apply to him because he is unprejudiced and can see things objectively. When another player disagrees with any of his views, the first player triumphantly shows that the opinion of his opponent can be ignored as the mere expression of class interest.

*Choice in education is only a device by which the middle classes can buy advantage for their children.*

(There is no point now in pointing to any role which competition might play in improving standards, or to the advantages of allowing parents some say in the type of education given to their children. You have already been convicted of trying to buy advantage; the rest is just cover.)

Skilful use of poisoning the well should employ both of its main characteristics. The poison should not only incite ridicule from the audience, it should also act as a deterrent to anyone tempted to disagree with you. 'Only an idiot' will put off some, but there will be others who think they could shrug it off. A better poison would be one sufficiently dreadful or embarrassing to deter anyone from drinking willingly.

*Only those who are sexually inadequate themselves now advocate single-sex teaching in our schools.*

(Any volunteers?)

Well-poisoning is recommended whenever your claim might not survive sustained scrutiny. It is also useful for dealing with an opponent whose point goes against received opinion but is, unfortunately, valid. Judicious poisoning will make such an opponent look so foolish that people will ignore the validity. It will also make you look witty and confident, and may even serve to conceal the fact that you are wrong.

# *Populum, argumentum ad*

The *argumentum ad populum* appeals to popular attitudes instead of presenting relevant material. In other words, it is based on prejudice. It exploits the known propensity of people to accept that which fits in comfortably with their preconceptions. The popular prejudices may or may not be justified, but the speaker who makes his case depend solely upon them is guilty of an *ad populum* fallacy.

> *In recommending Higginbottom, I'd point out that the smart money is on him.*

(Few people think they belong with the stupid guys.)

The *ad populum* is often equated with mob appeal, with inflaming passions and prejudices more appropriate to mass hysteria than to rational discourse. Mob orators make a career of the *ad populum,* choosing words calculated to raise the emotional temperature.

> *Are we to see the streets of this ancient land of ours given over to strange faces?*

(The prejudice is xenophobia and the implication is that the 'strange faces' do not fit in our streets; but no argument is advanced.)

Those who commit the fallacy take the easy way out. Instead of building up a case which carries conviction, they resort to playing on the emotions of the multitude. This is not sound logic, although it may be very successful. Conceivably Mark Anthony might have developed a case for punishing Brutus and the other assassins, and restoring Caesar's system of government. What he did was more effective. By appealing to popular rejection of disloyalty and ingratitude, and to popular support for public benefactors, he turned a funeral crowd into a rampaging mob.

For several centuries the traditional villains of the *ad populum* appeal were landlords and corn merchants. Although they play a negligible role in society nowadays, so powerful was their hold on popular prejudice that I expect you could still raise a lusty cheer by castigating opponents as profiteering landlords and corn merchants. Their disappearance has left a gap in the *ad populum* only partly filled by the mysterious 'speculators'. They are somewhat more nebulous because whereas letting property and dealing in corn were respectable occupations which could be identified, few people would write 'speculator' in the space for their occupation. Still, their elusiveness imparts a shadowy and sinister quality to enhance their evil.

> *I oppose enterprise zones because they will become sleazy red-light areas, characterized by sharp dealers and speculators.*

> (You have to be careful, though. Some audiences would like the sound of this.)

Your own *ad populums* will come naturally, since you are basically in support of the little man, the underdog, the local boy. The people against you are the big bosses, the money-men of the financial district and the bureaucrats on their index-linked pensions. 'Rich bankers' has lost its effect these days; most people equate it with their local bank manager who is not all that rich. Remember to use code-words where people feel the prejudice is not respectable. Racial minorities, for example, should be referred to as 'newcomers' or 'strangers', even when they have been here longer than you have.

*If we allow the corner shop to close, it will mean hard-earned money going out of the community to rich businessmen in flash cars. The corner shop is part of our locality; it's a friendly presence in the neighbourhood; it's the focal point of the community we grew up in.*

(People will do anything for it, except shop there.)

# Positive conclusion from negative premise

An argument which draws a conclusion from two premises is not allowed to have two negative premises, but it is allowed one, provided the conclusion is also negative. A fallacy is committed whenever a positive conclusion follows from two premises which include a negative one.

*Some cats are not stupid, and all cats are animals, so some animals are stupid.*

(Even though some of them are smart enough not to be cats, the conclusion does not follow. One premise is negative, so any valid conclusion would also have to be so.)

Although two things can be related to each other by means of the relationship which each has with a third, if one of the relationships is concerned with what is not true for one, the deduction must show that the other one is also wholly or partly excluded from some class. In other words, if they each enjoy a different relationship with a third thing, they cannot both be in the same class. The fallacy of drawing a positive conclusion from negative premises persuades us that things do belong to a class by telling of things which do not.

The trouble with this fallacy is that it can be seen coming a mile away. You can try persuading an audience that rats are sheep by telling them what rats are and what sheep are not. You are unlikely to succeed for the simple reason that people smell the rat before the wool is pulled over their eyes. It is just too easy to spot that you cannot claim that things are the same simply because they are different.

The only time you stand a chance of getting away with this one is when you are calling up a radio phone-in show. And that is only because anything goes on a radio phone-in show.

# *Post hoc ergo propter hoc*

The Latin translates as 'after this, therefore on account of this', and it is the fallacy of supposing that because one event follows another, then the second has been caused by the first.

*Immediately after the introduction of canned peas, the illegitimate birthrate shot up to a new high from which it did not decline until frozen peas edged canned peas out of the market. The link is all too obvious.*

(Too obvious to be true, perhaps. If your thoughts turn to feeding your daughters beans instead, remember to keep them clear of everything else which preceded the rise in illegitimacy. They should stay away from television, jet aircraft, polythene and chewing-gum, to name but a few of the more obvious hazards.)

Although two events might be consecutive, we cannot simply assume that the one would not have occurred without the other. The second might have happened anyway. The two events might both be linked by a factor common to both. Increased prosperity might influence our propensity to consume canned peas, and also to engage in activities which increase the rate of illegitimacy. Small children at gaming machines provide vivid illustrations of the *post hoc* fallacy. They may often be seen with crossed fingers, eyes closed, hopping on one leg, or in whatever physical contortion once preceded a win. They link their random preparations with the outcome of their luck; and in this they differ in no wise from more adult gamblers, whose concealed rabbits' feet and clenched-teeth incantations betray the same supposition. If it worked once, it can work again.

Unfortunately for our predictive ability, every event is preceded by an infinite number of other events. Before we can assign the idea of cause, we need rather more than simple succession in time. The philosopher David Hume pointed to regularity as the chief requirement, with some contiguity in time and space. We are more likely to describe a germ as the cause of an infection in a man if its presence has regularly preceded the infection, and if it is found in the body which is infected.

The charm of the *post hoc* fallacy emerges when we leave behind the everyday idea of cause and effect. Although we suppose we understand the mechanisms by which one event leads to another, Hume showed that it boils down to our expectation of regularity. The candle flame on the finger and the subsequent pain are called cause and effect because we expect the one to follow on regularly from the other. Of course, we concoct all kinds of explanations as invisible threads to link the two, but they come down to interposing unseen events between our first and second. How do we know that these unseen events really are the cause? Easy. They always follow from one another.

This gap in our knowledge provides a vacant lot in which fallacies can park at will. Greek historians regularly discussed natural disasters in terms of human actions. In looking for the cause of an earthquake, for example, we are likely to find Herodotus, or even Thucydides, gravely discussing the events which preceded it before concluding that a massacre perpetrated by the inhabitants of the stricken town was probably the cause.

The determined fallacist will see this as a field of opportunity. Whatever your opponent is urging is bound to have been tried somewhere, in some form, at some time. All you need do is attribute the unpleasant things which followed it to the operation of that factor. We know that unpleasant things followed it because unpleasant things are happening all the time; there are always plenty of earthquakes, sex offences and political broadcasts on TV which you can lay at your adversary's door.

> *'Imprisonment is barbaric. We should try to understand criminals and cure them using open prisons and occupational therapy.'*
> *'They have been trying that in Sweden since 1955, and look what's happened: suicides, moral degeneracy and drunks everywhere. Do we want that here?'*

(A term such as 'moral degeneracy' is the hallmark of the sterling fallacist, being more or less impossible to disprove.)

# Presentation over content

The way that an argument or a claim is presented does not alter its truth content or its logical validity, but it can often influence a listener or reader. A skilful, appealing presentation can make listeners or readers more inclined to accept the truth of what is being communicated more readily than might a poorer or more mundane presentation. The fallacy lies in the fact that the presentation is not part of the content and does not contribute to it. It is akin to the attractive packaging of a product that makes the customer more likely to choose it.

Just as the *ad hominem circumstantial* tailors the argument to the special interests of the audience, so the *presentation over content* fallacy tailors the way it is presented to appeal to their tastes and to the things that appeal to them, perhaps unconsciously.

The executive who puts his or her case to the board not as a collection of papers but bound into a book with a spine is adding nothing to the content of the proposal. However, he or she is drawing on the knowledge that many people take a book more seriously than a collection of papers. It seems more solid, more permanent.

The fallacy is a commonplace in the election manifestos of political parties. The manifesto content, if any, is what the party wants to do and feels it might be able to do. Focus groups and opinion polls are used to determine how those policies can be presented in the ways most likely to make them acceptable. This is not just a question of colours, layout, typeface and photographs, but of intangibles such as the feelings a particular presentation will arouse in the mind of the readers, if any.

> *The four promises we guarantee to the British people, set out clearly on coffee mugs to remind you every day of the firm pledges our party had made . . .*

(Two of the mugs arrived already broken.)

Of course there is a discipline in which presentation takes priority over content, and in which skilled practitioners go to enormous lengths to put things before the public in ways designed to secure an acceptance independent of the worth of what is being put across. But consideration of that field must await another time. This book is about logic, not marketing.

## *Quaternio terminorum*

*Quaternio terminorum* is the fallacy of four terms. The standard three-line argument requires that one term be repeated in the first two lines, and eliminated from the conclusion. This is because it works by relating two things to each other by first relating each of them to a third thing. This 'syllogistic' reasoning depends on one term, the 'middle term', being repeated in the premises but disappearing from the conclusion. Where there are instead four separate terms, we cannot validly draw the conclusion, and the *quaternio terminorum* is committed.

*John is to the right of Peter, and Peter is to the right of Paul, so John is to the right of Paul.*

(This looks reasonable, but one line has 'to the right of Peter' where the other one simply has 'Peter'. These are two separate terms, and the four-terms fallacy is involved. The conclusion is not validly established. After all, they could be sitting round a table.)

We might just as easily have said:

*John is in awe of Peter, and Peter is in awe of Paul, so John is in awe of Paul.*

(The error is more obvious. John might respect Peter for his intellect, and Peter could respect Paul for his Mercedes. Since John has a Bentley, he might not transfer his awe from Peter to the other cheap upstart.)

The fallacy arises because, strictly speaking, the terms in this type of argument are separated by the verb 'to be'. Whatever comes after it is the term. It can be 'the father of', or 'in debt to', or many other things. Unless the whole term appears in the next line, there is a *quaternio terminorum.* Of course, with four terms we cannot deduce new relationships between terms by using a middle term common to both – there isn't one.

*John is the father of Peter, and Peter is the father of Paul, so John is the father of Paul.*

(Even your grandfather can see this is wrong.)

Now look at the example where there is a middle term repeated:

*John is the father of Peter, and the father of Peter is the father of Paul, so John is the father of Paul.*

(There are three terms, and this is valid.)

*Quaternio terminorum* can result is endless confusion in daily relationships. If John is in debt to Peter to the tune of 45 dollars, and Peter is in debt to Paul (who saved him from drowning), John might be very surprised to find Paul on his doorstep demanding money with menaces. On the other hand, if John is in love with Mary, and Mary is in love with Paul, no one except a theatre dramatist would attempt to complete the fallacious deduction.

The four-terms fallacy is more likely to appear as a source of genuine error than of deliberate deception. People may fool themselves with arguments constructed around it, but they are unlikely to fool others. There is something about the odd look of it which alerts the unwary; it is like a cheque without the amount filled in. No date perhaps; maybe even no signature; but everyone looks at the amount.

> *China is peaceful towards France, and France is peaceful towards the USA, so China must be peaceful towards the USA.*

(You do not even need to know anything about China to know this is wrong. Just remember not to trust any relationship with France in it.)

One way to use the fallacy with a fair chance of success is to smuggle it in amongst a group of comparatives. Comparatives, such as 'bigger than', 'better than', 'stronger than', or 'fatter than', do work because they are transitive, despite the four terms. After a few of these, slip in the non-transitive relationship and it might get by.

> *Darling, I'm bigger than you are, stronger, and richer; yet I respect you. You stand in the same relationship to your mother, so I, in turn must respect your mother.*

(I just don't want her in the house.)

# The red herring

When the hounds are intent on following a scent of their own choosing, in preference to that selected by the master of the hunt, a red herring is used to effect the transfer. Tied to a length of string, it is led across the trail the hounds are following. Its powerful aroma is sufficient to make them forget what they were following, and to take up its trail instead. The red herring is then skilfully drawn onto the trail which the hunt-master prefers.

In logic the red herring is drawn across the trail of an argument. It is so smelly and so strong that the participants are led irresistibly in its wake, forgetting their original goal. The fallacy of the red herring is committed whenever irrelevant material is used to divert people away from the point being made, and to proceed towards a different conclusion.

> 'The police should stop environmental demonstrators from inconveniencing the general public. We pay our taxes.'
> 'Surely global meltdown is infinitely worse than a little inconvenience?'

(It may well be, but that particularly ripe and smelly fish is not the one we were following.)

The use of the red herring is fallacious because it uses irrelevant material to prevent a conclusion being reached in its absence. If the argument leads in a particular direction because reason and evidence are taking it there, it is not valid to divert it by means of extraneous material, however attractive that new material may be.

> 'Excuse me, sir. What are you doing with that diamond necklace hanging out of your pocket?'

*'I say, isn't that a purebred German shepherd dog you have with you?'*

(Even if the policeman is put off the scent, the dog won't be.)

The more the red herring appears to follow the original trail for a little way, the more attractive it is to follow, and the more effective it will be at diverting attention.

*'Publicans always try to promote whatever makes them most profit.'*

*'I think these fashions come and go. One time they will promote light beer because they think that is where the demand is; but a year or two later it might be cask-conditioned ale.'*

(The attraction here is that it smells a little like the original trail. It talks about what publicans promote, but after following this one for an hour or two, the talkers will be as much fuddled by the argument as by the beer.)

Red herrings are used by those who have a bad case, and can feel the hounds getting uncomfortably close to it. Politicians under pressure will toss out so tempting a red herring that the dogs will turn after it, even in the act of leaping for the kill. Lawyers scatter them at the feet of juries to divert attention away from crooked clients. Every famous attorney has been credited with the trick of putting a wire through his cigar so that, instead of listening to the details of his weak case, the jurors watch with bated breath as the ash grows longer and longer. The red herring in this case is a visual one, like the salesman's illuminated bow tie which diverts attention away from his inferior product.

*'You* never *remember my birthday.'*

*'Did I ever tell you what beautiful eyes you have?'*

You should never set out upon a weak argument without a pocketful of red herrings to sustain you through the course of it. As your intellectual energies begin to fail, your supply of them will give you breathing space. If you aspire to the ranks of the experts you should select your red herrings on the basis of the known interests of your audience. Every pack has its favourite aroma; and your red herrings should be chosen with that in mind. As you toss them out as needed, the audience will be unable to resist their favourite bait. You can gain respite in the most difficult situations by skilfully introducing the subject of the arguer's bad back, or even his summer holidays. In real desperation you can bring up his pet cat.

# Refuting the example

Examples are often adduced in support of an argument. When attention is focused on showing the example to be a false one, but leaving the central thesis unchallenged, the fallacy is known as 'refuting the example'.

> *'Teenagers are very bad-mannered these days. That boy from next door nearly knocked me over in the street yesterday, and didn't even stay to apologize.'*
> *'You're wrong. Simon is no longer a teenager.'*

(None of which knocks over the original assertion, only one example.)

While an example can illustrate and reinforce an argument, the discrediting of it does not discredit the argument itself. There may be many other instances which support the thesis, and which are genuine cases.

There is a fine distinction between the quite legitimate activity of casting doubt on an opponent's evidence, and concentrating criticism on the example instead of on the thesis which it supports. If the rejection of the central claim is urged only because a bad example was used to support it, the fallacy is committed.

> *I can show that there is no truth at all in the allegation that hunting is cruel to animals. In the case of the Berkshire hunt described to us, what we were not told was that a post-mortem showed that this particular fox had died of natural causes. So much for charges of cruelty.*

(The argument has less life in it than the fox.)

A case of this fallacy occurred in a general election. One party featured a poster showing a happy family to illustrate the slogan that life was better with them. Their opponents devoted an extraordinary amount of time and attention to the actual model who appeared in the photograph, and to publicizing the fact that his was not a happy marriage. The effort was presumably expended in the belief that the public would be less likely to believe the fact once the example was refuted.

For some reason this fallacy is very prevalent in discussion about sport. In support of a generalized claim, such as 'Spain produces the best strikers', an example will be produced. This seems to be the cue for lengthy and dull evaluation of the merits of the individual concerned. The assumption throughout the discussion seems to be that the case for or against the original general statement will be won or lost with that of the example.

You can set up situations for using this fallacy by prodding your opponents with a demand for examples. Your heavy scepticism as you respond to their claims with 'such as?' will prompt them into bringing forward a case in point. Immediately they do so, you attack the case, showing how it could not possibly be valid. The case of a family produced to show that bus drivers' pay is too low can be attacked with great merriment by asking whether they have a colour television and how much the husband spends on beer. Even if you cannot undermine the example in apparent destruction of the assertion it supports, you can probably widen the talk to a more general discussion about what constitutes poverty, and cast doubt on whether the original statement means anything at all. This is called 'linguistic analysis'.

# Reification

The fallacy of reification, also called hypostatization, consists in the supposition that words must denote real things. Because we can admire the redness of a sunset, we must not be led by the existence of the word into supposing that redness is a thing. When we see a red ball, a red table, a red pen and a red hat, we commit the fallacy of reification if we suppose that a fifth object, redness, is present along with the ball, the table, the pen and the hat.

> *In SKYROS we have extracted the blueness of the summer sky and inserted it in a bar of heavenly soap.*

> (Since the 'blueness' of the summer sky is not an object, it cannot be processed like a material thing.)

Turning descriptive qualities into things is only one form of reification. We can also make the mistake of supposing that abstract nouns are real objects.

> *He realized that he had thrown away his future, and spent the rest of the afternoon trying to find it again.*

> (If you think that sounds silly, you should watch Plato searching for justice.)

Sometimes objects have consequential attributes, in their arrangement, perhaps. We commit reification if we suppose that these attributes are as real as the objects they depend upon.

*It* [the Cheshire Cat] *vanished quite slowly, beginning with the end of the tail, and ending with the grin, which remained some time after the rest of it had gone.*

(Alice could see it because she had sharp eyes. After all, she had seen nobody on the road, while the Duchess had difficulty enough seeing *some*body.)

The fallacy occurs because our words have not the power to conjure up real existence. We can talk about things which do not exist at all, and we can talk of things in one form which actually might exist in another. 'Redness entered the sky' says roughly the same as 'the sky reddened', but the words denote different activities. Our words are not evidence for the existence of things; they are devices for talking about what we experience.

There is a school of philosophers which believes that if we can talk about things they must, in a sense, exist. Because we can make sentences about unicorns and the present king of France, they claim that there must actually be unicorns and a present king of France (with the latter presumably riding on the back of the former).

Yet another school elevates the fallacy into an art form, by talking about the 'essences' of things. They claim that what makes an egg into an egg and nothing else is its 'eggness', or the essence of egg. This essence is more real and more durable than the actual egg, for ordinary eggs disappear into quiche lorraine, but the idea of an egg goes on. The obvious objection, that this is just silly, is a commanding one. We use words like labels, to tie onto things so we do not have to keep pointing at them and communicating in sign language. Little can be inferred from this except that we have agreed to use words in a certain way. If someone brings out the 'essences' behind your words to show you what you really believe in, change the words.

*'You claim to support freedom, but the whole liberal democratic system has the essence of slavery.'*

*'All right. We'll call it slavery, then. And let it be clear that by "slavery" we mean people voting as they wish in elections, having a free press and an independent judiciary, etc.'*

(This is an upsetting tactic. The accuser had expected the old image of slaves being whipped on plantations to be carried over into the new use describing the Western democracies.)

Your own use of reification can be directed toward showing that what people say they support involves them in supporting your position. You simply take all of the abstract concepts, turn them into real entities, and start demonstrating that their *real* natures are in line with what you were saying.

*You say that God exists, but let us look at this idea of existence. We can talk about tables which have existence, chairs which have existence and so on, but for* pure *existence you have to take away the tables and chairs and all of the* things *which exist, to be left with existence itself. In taking away everything which exists, you are left with nothing existing, so you see the existence of your God is the same as non-existence.*

(He'll never spot that existence doesn't exist. After all, Hegel didn't.)

# The runaway train

A runaway train takes you speeding into the distance, but unfortunately does not stop. This means that when you reach your required destination you cannot leave it, but are compelled to be taken further than you wished. The runaway-train fallacy is committed when an argument used to support a course of action would also support more of it. If you wish to stop at a particular point, you need an argument to do so.

It might well be true that lowering a highway speed-limit from 70 mph to 60 mph would save lives. That is not a sufficient argument for choosing 60 mph, however, because lowering the speed-limit to 50 mph would save even more lives. And more still would be saved at 40 mph. The obvious conclusion of this runaway train is that if saving lives is the sole aim, the speed-limit should be set at the level which saves the most, and this is 0 mph.

In practice the lives at risk for each proposed speed-limit have to be measured against what is achieved by the ability to travel and to transport goods rapidly. Most of our daily activities involve a degree of risk which could be reduced if we limited our actions. In practice we trade off risks against convenience and comfort. If the case for making the speed-limit 60 mph is based solely on the lives which could be saved, the arguer will need additional reasons to stop at 60 mph before the *runaway train* of his own argument takes him to 50 mph, then 40 mph, and finally crashes into the buffers when it reaches 0 mph.

People argue that since, in the UK, everyone has to pay for the country's National Health Service, this gives the state a sufficient justification to ban smoking, because smokers suffer more illnesses. There may be good reasons to ban smoking, but the argument that the costs of the smoker's behaviour should be imposed on others is a runaway train. Why stop there? The same argument applies to all behaviour which affects health adversely. It could be applied to the eating of saturated fats such as butter, or refined white sugar. The state could require people to exercise in order to prevent the health costs of their laziness from falling on others. If this argument is to apply only to smoking, there have to be reasons why the train stops there.

Someone boards a runaway train when they are so concerned about direction that they forget to attend to distance. They can continue happily on their journey until their reverie is broken by someone calling 'Why stop there?'

*The state should subsidize opera because it would be too expensive to mount productions without the extra support from public funds.*

(And as the train heads off into the distance, wait for the stations marked *son et lumière* concerts, civil war re-enactments, and gladiatorial displays. If opera is different, we need to know why.)

The fallacy is often committed when someone advances a general argument for something he regards as a special case. If the argument has any merit, the listener immediately wonders why it should be limited to that case. To combat a runaway train, it is usually sufficient to point to some of the absurd stations further down the same line. If good schools are to be banned because they give children an 'unfair' advantage, why not prevent rich parents from doing the same by buying their children books, or taking them on foreign holidays?

To lure people on board the runaway train, simply appeal to things which most people favour, like saving lives, aiding widows and orphans and having better-behaved children. Use the general support which such things enjoy to urge support for the one proposal you favour which might help to achieve them, even as you carefully ignore the others.

In a very specialized use of the fallacy, you should gain acceptance of the principles to support a reasonable objective, and only when that point has been reached, reveal the unreasonable objective also supported by the same principles.

*You agreed to allow a bingo hall in the town because people should have the choice to gamble if they want to. I'm now proposing to have gaming machines on every street corner for precisely the same reasons.*

# Secundum quid

The fallacy of *secundum quid* is otherwise known as the hasty generalization. Whenever a generalization is reached on the basis of a very few and possibly unrepresentative cases, the fallacy is committed. It takes the argument from particular cases to a general rule on the basis of inadequate evidence.

*I was in Cambridge for ten minutes and I met three people, all drunk. The whole place must be in a state of perpetual inebriation.*

(Not necessarily so. Saturday night outside Trinity College might be quite different from King's on Sunday. A similar conclusion about London might have been drawn by a visitor who saw three people at midday outside a newspaper office.)

The fallacy lies in the assumption of material which ought to be established. There should be an attempt to establish that the sample is sufficiently large and sufficiently representative. One or two cases in particular circumstances do not justify the presumption of a general rule, any more than the sight of a penny coming down heads can justify a claim that it will always do so.

Behind our identification of the fallacy lies our recognition that the few cases observed might be exceptional to any general rule which prevails.

*Don't shop there. I once bought some cheese and it was mouldy.*

(This smells like a broad condemnation placed on a narrow base.)

Clearly there is fine judgement required to distinguish between a *secundum quid* and a case where one or two instances do enable a valid judgement to be made. When assessing the fitness of a candidate for foster-parent, for example, it would be prudent to make a judgement on the basis of only one previous incident of child-molesting. In the film *Dr Strangelove,* when a psychotic commander sends his wing on a nuclear attack against the USSR, the General reassures the President: 'You can't condemn the whole system just because of one let-down.' Both of these cases deal with systems which seek 100 per cent safety coverage, and in which one exception *does* validate a judgement. *Secundum quid* covers the more general circumstance in which it does not.

A visitor who assesses the population of London from his experience of a royal wedding day is likely to be as wrong as one who makes a similar judgement about Aberdeen on a charity-collection day. The basic rule is 'don't jump to conclusions'.

Opinion pollsters try to be very careful to avoid *secundum quids.* A famous American poll once wrongly predicted a Republican victory because it surveyed by telephone, not realizing that fewer Democrats owned telephones. Political parties everywhere are not averse to 'talking up' their support by quoting obviously unrepresentative poll-findings.

Scientific knowledge is like a battlefield mined with *secundum quids.* Scientific theories are often put forward with only a very few examples to back them up. The problem is one of knowing when there are enough case-histories to be sure about the general rule put forward to explain them. Astonishingly, the answer is never. Science proceeds with the knowledge that a new case could suddenly appear to show that even its most solid theories are no good. A billion apples might have hit a billion heads since Newton's, but it would still take only one apple going upward to force at least a modification to the general theory.

*Secundum quids* will be very useful to you in persuading audiences to pass judgements which coincide with your own. You should appeal to one or two cases, well-known ones if possible, as proof of a general judgement.

*All actors are left-wing subversives. Let me give you a couple of examples . . .*

(You then spread over the entire profession the tar which your brush collected from two of them.)

# Shifting ground

People may employ hedging to make their contentions ambiguous, or they may use a definitional retreat to claim that their words meant something else. In the third version of this defensive type of operation, they may actually change the whole ground they were maintaining, while still claiming continuity. When people do shift the substance of what they were saying, they commit the fallacy of shifting ground.

> *I said I liked the project and thought it a good one. However, I share the objections you have all voiced, and can only say how much this reinforces a view I have long held that it is not enough for a project to be likeable and good.*

(A leap from one bank to the other with the grace of a ballet dancer superimposed on the desperation of a stranded man.)

The deception is the source of the fallacy. Criticism of the original stance is avoided by shifting to a different one. In that the argument has taken place about the position as understood, it is irrelevant to the new position which is now claimed. Similarly a critique now has to start all over again on the new position because what has passed so far has not been centred upon it.

> *I said we'd come out stronger after this election. Look, we both know that many things can strengthen a party. I have always thought it a source of strength if a party can respond to criticism. Now, with our share of the poll down to 9 per cent, I think that . . .'*

(This can be seen in every election by every party except the winner. It is roughly equivalent to 'I don't think that a score of five goals to one against us should be seen as a defeat for Scottish football. It is more of a challenge which . . .')

The shifting sands of political fortune often coincide with the shifting ground of the fallacy. This is because of a patently absurd rule that no politician must ever change his mind about anything. To do so would be to admit he was wrong before, and could, by implication, be wrong now. Infallibility must, therefore, be sustained. Shifting ground, insecure though it might look to us, provides a solid foundation for political continuity.

There is a certain class of religious argument in which anything at all whose existence is assented to can be called divine. Here the base of discussion seems to slide quite happily across several continents, as what started out as a discussion about a man in the sky with a white beard ends up in consideration of some abstract principle of the universe.

Shifting ground is for defensive use. You cannot convince others of a new point with it, but you can use it to avoid it being known that you were wrong. As the victorious armies march into your territory after the struggle, they are surprised to find you at the head of them, leading the invasion. They had quite mistakenly supposed that you were head of the defence forces.

*After hearing his point of view, I feel that Mr Smith's amendment to insert the word 'not' into my motion expresses the spirit of what I was trying to say. I therefore accept his amendment as an improvement to my motion.*

There are muscular exercises which you should practise every day in front of a mirror, and which assist the mental contortions needed to shift ground rapidly.

*Yes, I walked through the green line, customs officer, and I can explain that extra bottle of scotch.*

(Does anyone spot the slight tremor in his feet?)

# Shifting the burden of proof

Shifting the burden of proof is a specialized form of the *argumentum ad ignorantiam.* It consists of putting forward an assertion without justification, on the basis that the audience must disprove it if it is to be rejected.

Normally we take it that the new position must have supporting evidence or reason adduced in its favour by the person who introduces it. When we are required instead to produce arguments against it, he commits the fallacy of shifting the burden of proof.

> *'Schoolchildren should be given a major say in the hiring of their teachers.'*
> *'Why should they?'*
> *'Give me one good reason why they should not.'*

(It always looks more reasonable than it is. You could equally ask that the janitor, the dinner-ladies and the local bookie be given a say. Come to think of it, they might do a better job.)

It is the proposal itself which has to be justified, not the resistance to it. The source of the fallacy is the implicit presumption that something is acceptable unless it is proved otherwise. In fact the onus is upon the person who wishes to change the status quo to supply reasons. He has to show why our present practices and beliefs are somehow inadequate, and why his proposals would be superior.

> *I believe that a secret conspiracy of* Illuminati *has clandestinely directed world events for several hundred years. Prove to me that it isn't so.*

(We don't have to, anymore than we have to prove that it isn't done by invisible elves or Andromedans living in pyramids under the Bermuda triangle.)

The maxim of William of Occam, usually shortened to 'entities should not be multiplied beyond necessity', tells us not to introduce more by way of explanation than is needed to explain. World events are already explained by divine purpose, evolutionary progress or sheer random chaos. We do not need *Illuminati* added to the brew, and he who would introduce them must show what evidence requires them to explain it.

Shifting the burden is a very widespread and common fallacy. Popular conception has it that he who says 'prove it' and he who says 'prove it isn't' are on equal ground. It is a misconception. The one who asks for proof is simply declaring an intention not to accept more than the evidence requires. The other is declaring his intent to assume more than that.

This particular fallacy is the frail prop on which rests the entire weight of unidentified flying objects, extrasensory perception, monsters, demons and bending spoons. Advocates of these, and many other, ethereal phenomena try to make us accept the burden of establishing falsity. That burden, once taken up, would be infinite. Not only is it extraordinarily difficult to show that something does *not* exist, but there is also an infinite load of possibilities to test.

You will need shifting the burden of proof if you intend to foray into the world of metaphysical entities. Instead of resorting to the simple 'you prove it isn't', you should clothe your fallacy in more circumlocutious form,

*Can you show me one convincing piece of evidence which actually disproves that . . .?*

(This tempts the audience into supplying instances, giving you a chance to slide into 'refuting the example' instead of giving any arguments in favour of your case.)

The popular misconception about the onus of proof will enable you to put forward views for which there is not a shred of evidence. You can back gryphons, the perfectibility of man, or the peaceful intentions of religious fundamentalists.

# The slippery slope

Slippery slopes are so tricky to negotiate that even the first timid step upon them sets you sliding all the way to the bottom. No one ever goes up a slippery slope; they are strictly for the descent to disaster. The fallacy is that of supposing that a single step in a particular direction must inevitably and irresistibly lead to the whole distance being covered. There are cases in which one step leads to another, and cases where it does not. It is not a fallacy to suppose that after the first stride, further steps might be taken towards unpleasant consequences, but it is usually an error to suppose that they must.

There is a limited class of cases in which someone is doomed after the first step; stepping off a skyscraper is one of them. But in most life situations there is a choice about whether or not to go further. Those who oppose progress, however, often use the slippery-slope argument to suggest that any reform will lead inexorably to unacceptable results.

*I oppose lowering the drinking age from 21 to 18. This will only lead to further demands to lower it to 16. Then it will be 14, and before we know it our new-borns will be suckled on wine rather than mother's milk.*

The point is that the factors which lead to the arbitrary drinking age of 21 might change. There is nothing which suggests that they must keep on changing, or that society must keep on responding.

The slippery slope basically argues that you cannot do anything without going too far. This belies human progress, which has often been made by taking short steps successfully where longer ones might have been ruinous.

*If we allow French ideas on food to influence us, we'll soon be eating nothing but snails and garlic and teaching our children to sing the* Marseillaise.

(It might beat pizza and chips, though.)

In some cases there is a point of principle at stake which, once yielded, allows anything. This is not so much a slippery slope, however, as a vertical drop. The story is told of a dinner-table conversation between the dramatist George Bernard Shaw and a pretty lady:

> *'Would you sleep with me for a million pounds?'*
> *'Why yes, I would.'*
> *'Here's five pounds, then.'*
> *'Five pounds! What do you think I am?'*
> *'We've established that. Now we're talking about price.'*

(Shaw was correct, but this is not a slippery-slope argument which would have led the lady to immorality in stages. Once the principle was conceded, the rest was bargaining.)

On a slippery slope ruin is reached in stages. The fallacy introduces the irrelevant material of the consequences of more far-reaching action in order to oppose the more limited proposal actually made.

Use the fallacy yourself to oppose change. There is scarcely any proposal which would not lead to disaster if taken too far. They want to charge people for admission to the church bazaar, but you point out that if this is conceded they will charge more next year, and more after that, until poorer people will be unable to afford to get in. The fallacy works best on pessimists, who are always ready to believe that things will turn out for the worse. Just assure them that if they do anything at all, this is almost certain to happen.

# Special pleading

Special pleading involves the application of a double standard. Although the normal rules of evidence and argument are applied to other cases, the fallacy of special pleading stipulates that some are exceptions, to be judged differently. It normally occurs when a speaker demands less strict treatment for the cause which he espouses than he seeks to apply elsewhere.

*Our attempt to engage in conversation was totally spoiled by all the chattering that other people were doing.*

(Look who's talking.)

Special pleading is a source of error. If different standards are to be applied to certain cases, we need rather more evidence to justify this than the fact that we would like better treatment. The same standards which would throw out someone else's claim will also throw out our own. If we were to receive special treatment, how could we justify withholding it from others? Argument proceeds by general rules, and exceptions must be justified.

*While it is not normally right to invade someone's privacy, it is all right for us, as journalists, to do so because we serve a public need.*

(Even though we make private money.)

Special pleading is sometimes described as 'benefit of clergy', because of the right which the medieval Church established to have clerical offenders tried in church courts even for civil crimes. This right, which was called 'benefit of clergy', is really what the special pleader seeks – the right to be tried in a different court.

*Capitalism has always left areas of poverty and hardship, and mis-allocated resources. Socialism, on the other hand, has never been properly tried.*

(Can you spot the special pleading? We are invited to compare capitalism in practice, as applied, with theoretical socialism. This is sometimes called 'real' socialism, to conceal the fact that it is the opposite of real. Of course, if we look at capitalist countries for the record of capitalism, then we should look at socialist countries for the record of socialism. Theory with theory, or practice with practice.)

Special pleading is normally resorted to by those whose case would not fare well in the general courts. Faced with a clash between their ideas and the evidence, scientists change their ideas. The special pleaders, like social scientists, prefer to change the evidence and show why normal judgements cannot be made in their particular case. Very often it is the supreme importance of the cause which is called upon to justify the special standards.

> *Normally I would object to spitting at public figures, but the threat of global warming is so awful . . .*

(As is the threat of fluoridation, Sunday trading and canine nudity. It depends on how strongly you feel about it.)

On a personal level, we are all apt to be more tolerant of ourselves than we are of others. For behaviour we would universally condemn from others, we invent excuses to forgive it in ourselves. Our queue-jumping is excused by urgency, but not that of anyone else. Our impulse buying is justified by need; others who do it are spendthrift. The same standards which excuse ourselves also excuse our team, our group, our town and our country.

When using special pleading in support of your own side, take care that you always supply some specious justification to account for the exception from the general rule. It is never just because it is your side which is involved; always there are special circumstances of public interest.

> *With any other boy I'd be the first to admit that burning down the school was wrong, but Michael is very highly strung, as talented people tend to be . . .*

(Talented people get away with arson, it seems, as well as murder.)

# The straw man

The straw man of logic does not scare anyone. No self-respecting crow would even rustle a feather at him; he is too easy to knock down. Precisely. The straw man is made incredibly easy to knock down so that when you are unable to refute your opponent's argument, you can topple the straw man instead. The straw man is, in short, a misrepresentation of your opponent's position, created by you for the express purpose of being knocked down.

> *We should liberalize the laws on marijuana.*
> *'No. Any society with unrestricted access to drugs loses its work ethic and goes only for immediate gratification.'*

(Down he goes! The proposal was to liberalize marijuana laws, but 'unrestricted access to drugs' makes a much less stable target.)

Traditionally, the straw man is set up as a deliberate overstatement of an opponent's position. Many views are easier to argue against if they are taken to extremes. If your opponent will not make himself an extremist, you can oblige with a straw man. Any easily opposed misrepresentation will serve as your dummy.

The straw man is fallacious because he says nothing about the real argument. Like the *ignorati elenchi* society he belongs to, he is totally beside the point. His function is to elicit, by the ease of his demolition, a scorn which can be directed at the real figure he represents.

Aficionados of the straw man ploy reserve their loudest *olés* for those whose straw construction is concealed by a layer of flesh. The point is that the straw man does not always have to be created specially. By deliberately picking on a weak or absurd supporter of the opposition, and choosing to refute him instead of the main protagonist, you indulge in the true connoisseur's use of the straw man.

Even today, applause can be gained for 'refuting' the theory of evolution, so long as one is careful to refute Darwin. Modern evolutionary theory is more advanced, having knowledge of things such as genetics to help it along. But you can set up Darwin as a straw man and, by knocking him down, give the impression you have 'refuted' the theory of evolution.

It is standard practice at elections to choose the most foolish or ignorant spokesman for the other side to deal with, as well as to fabricate extremists who can be felled with a scornful half-line.

> How can we support the Democrats when one of their own union backers publicly advocates a 'worker state' like Soviet Russia was?

> (Biff! Bam! And another straw man bites the dust. Union leaders on one side, like businessmen on the other side, can be politically naïve, and make much better targets than the slippery eels who lead the parties.)

Historically, the role of the straw man has been to show the dangers of change. A handful of reformers or radicals advocating greater liberty or greater tolerance have been trampled to death by legion upon legion of straw men in serried ranks calling for anarchy, licence, the destruction of society and the slaughter of the innocents.

Use of the straw man is fun. Everyone needs a victory or two for purposes of morale. If real ones are nowhere to be had, then walloping the occasional straw man can be most invigorating. In addition to the advice already given, you would be wise to construct and demolish your straw man, wherever possible, after your opponent has uttered his last word on the subject. Your straw man looks pretty silly lying in the dust if your adversary is there to disown him. If your opponent is absent, or has finished his piece, there will be no one to deny that the crumpled figure lying at your feet is indeed the opponent you were facing, rather than a dried-grass dummy, hastily fabricated to take the fall in his place.

# *Temperantiam, argumentum ad*

If fallacies were assigned to the nations of the world, the *argumentum ad temperantiam* would be allocated to England. It is the Englishman's fallacy. The *argumentum ad temperantiam* suggests that the moderate view is the correct one, regardless of its other merits, it takes moderation to be a mark of the soundness of a position.

> *The unions have asked for 6 per cent, the management have offered 2 per cent. Couldn't we avoid all the hardship and waste of a lengthy strike, and agree on 4 per cent?*

> (If we did, next time the unions would demand 20 per cent and the management would offer minus 4 per cent.)

The *argumentum ad temperantiam* appeals to a common instinct that everything is all right in moderation. Moderate eating, moderate drinking and moderate pleasures have been widely praised by cloistered philosophers without any extreme desires of their own. The *ad temperantiam* appeals to that upper-class English feeling that any kind of enthusiasm is a mark of bad manners or bad breeding. One shouldn't be too keen. It helps to explain why none of them are particularly good at anything, and accounts for their steady, but moderate, decline.

The fallacy enters in because, while moderation may be a useful maxim to regulate our desires, it has no specific merit in argument. Where one view is correct, there is no rule that it will be found by taking the average or mean of all of the views expressed.

If two groups are locked in argument, one maintaining that $2 + 2 = 4$ and the other claiming that $2 + 2 = 6$, sure enough, an Englishman will walk in and settle on $2 + 2 = 5$, denouncing both groups as extremists. He is correct to describe them as extremists, but incorrect to suppose that this proves them wrong.

*I have tried, during my term of office, to steer a middle course between partiality on the one hand and impartiality on the other.*

(He might have added: between truth and falsehood, between vice and virtue, between falling asleep and staying awake, between sense and nonsense.)

In countries and situations where bargaining is more common than fixed price transactions, people routinely manipulate the extremes in order to influence the idea of a 'fair' average. Exactly the same procedure can be used in public life, advocating an extreme position in order to pull the eventual settlement closer to your way of thinking.

Only in England do people write books with titles like *The Middle Way,* elevating the *argumentum ad temperantiam* into a guide for public policy. The Liberal Party used to make a career of the fallacy, regularly taking up a position midway between those of the two main parties, and ritually denouncing them for extremism. The main parties, in their turn, contained this threat by bidding for 'the middle ground' themselves. This led the Liberals to become extremists in order to attract attention. In Britain New Labour was built upon the *temperantiam.* They called it the Third Way.

*One side represents capitalism; the other stands for socialism. We offer instead a policy of co-partnership to replace the old politics of conflict and extremism.*

(So alluring is this type of thing to the *ad temperantiam* mind, that the other parties hastily produce versions of it.)

When you use the *argumentum ad temperantiam* yourself, you should try to cultivate that air of smug righteousness which shows it to best advantage. Remember that your opponents are extremists, probably dangerous ones. They are divisive and destructive. Only you, taking the middle course, tread the virtuous path of moderation.

You will find it useful to invent extreme positions on one side, in order to cast the opposing views as extremist also.

*Councillor Watson has urged free travel for senior citizens. Others have suggested we should charge them 50 pence per journey. Surely the sensible course would be to reject these extremes and opt for a moderate charge of 25 pence?*

(Of course, the debate was between 25 pence and zero. The 50 pence advocates are conjured up in support of your *ad temperantiam*.)

Try to cultivate the company of Foreign Office officials. It comes so naturally to them when someone makes a claim against Britain to concede half of it that you will learn to commit the fallacy at speed with apparent ease. You will need to be quick off the mark because the fallacy has a large following.

When two countries are disputing the ownership of a couple of islands for example, you should be the first to leap in with the 'one each' suggestion. There will be plenty of British diplomats trying to beat you to it.

# Thatcher's blame

When the round black hat first appeared it was dubbed a bowler. This was because it looked like a bowl, and because it was made by the Bowler brothers. The term 'Thatcher's blame' might similarly catch on for two reasons: it was regularly used against the lady herself, and it covers all cases, just as a thatcher covers all of a roof.

In her first few years in office, Lady Thatcher was blamed for poverty and unemployment in Britain. Seamlessly this switched to blame for the culture of shameless affluence as the emerging class of yuppies flaunted their new-found wealth. She was deemed to be at fault in both cases.

The fallacy of 'Thatcher's blame' is committed when blame is attached no matter what outcome ensues. The fallacy occurs because the evidence is irrelevant when the determination of guilt precedes the outcome of their actions. Indeed, the point about 'Thatcher's blame' is that it covers all the conceivable outcomes.

If a policy is introduced first in Scotland, ahead of its application in England, the accusation is that the Scots are being used as guinea-pigs, and put at risk simply to test it. On the other hand, if the policy is introduced in England *before* being extended to Scotland, the charge will be made that the Scots are being left out yet again. Finally, if the policy is introduced at exactly the same time in both countries, this will be taken as evidence that the policy-makers are failing to appreciate the essential differences between England and Scotland. Heads you lose, tails you lose, and if the coin lands on its edge you also lose.

The fallacy works well in parliament because the official opposition is supposed to oppose. 'Thatcher's blame' allows them to be against whatever the government decides to do, no matter what the outcome might be. Thus anything done quickly is being 'rushed through recklessly', while measures which take time are tagged with 'intolerable delays'.

The fallacy falsely pretends that a judgement is being made based on the outcome, when that negative judgement would have been applied to any outcome. It regularly appears in Britain's tabloid press, where once a celebrity has fallen from favour any action they take is deemed to deserve condemnation. Since the opprobrium comes anyway, it expresses no real judgement on the morality or merits of the actions themselves.

> *I've been asked to a christening, but I'm sure they'll give the child some outlandish name that will make it a laughing-stock. Either that or some unbelievably tedious and commonplace name which will make the child seem like a faceless conformist.*

The fallacy is easy to use because it preys upon an instinct which would rather hear ill than good about people. After all, gossips don't go round praising people for their righteous actions. To use it effectively, you should pour scorn on some proposed action, predicting an adverse outcome. You then introduce an alternative consequence with the words 'And even if . . .' This allows you to predict more dire consequences. Your audience will never spot that you have, like the fallacy, covered every conceivable case. If you think this too obvious, reflect that for over a century the followers of Marxism predicted disaster for capitalism, whatever outcomes it produced.

# Trivial objections

The problem with trivial objections is that they leave the central thesis largely untouched. It is fallacious to oppose a contention on the basis of minor and incidental aspects, rather than giving an answer to the main claim which it makes.

> *I am totally opposed to the new road around the town. It will make all of our town maps out of date.*

> (It is rare for the fate of a new road to be decided on the basis of what it does to the maps. That said, however, one cannot help noticing that the maps show that towns reach very strange decisions on such matters.)

The fallacy is akin to that of the straw man. Instead of facing the main opponent, in this case it is only a few aspects of it which are confronted. The trivial objections are possibly valid; the point is that they are also trivial, and not adequate to the work of demolishing the case which is presented. The fallacy is committed because they are not up to the task to which they are assigned, not because they are erroneous.

*We cannot countenance any involvement in a land war in Europe. Think of what it might do to the supply of long-life milk from the continent.*

(Integrity, honour and glory sometimes seem pretty trivial reasons – but long-life milk . . .)

Associate membership of the European Union, when it was known as the European Economic Community, was, however, rejected by a British prime minister as 'beneath our dignity'.

Trivial objections tend to appear when the central thrust of the argument is difficult to oppose. Very often they make their appearance as practical difficulties put in the way of a popular proposal.

*Although banning cars from the High Street will severely hit trade at my own store, I would still go along with the majority but for one thing. We do not have a single sign-writer in the area who could make up the necessary road signs.*

It is often difficult to oppose the democratic process without appearing to be undemocratic. The fallacy of trivial objections permits a combination of readiness to accept the idea with hostility to any practical proposal. Elections can be opposed because of the paperwork involved. Referenda, while good in principle, can be opposed on grounds of cost.

*Of course we, as teachers, would like the parents to have the final say on this; but there just isn't a hall big enough for such a meeting.*

(A meeting of teachers who really favoured the proposal could meanwhile be held in the store-cupboard.)

When you are searching for trivial objections with which to do down ideas which are difficult to oppose head-on, you can always drag up objections from highly unlikely hypothetical situations.

*Yes, vicar, I would like to come to church more regularly. But suppose the house caught fire one Sunday morning while I was away?*

(Why, it would then become another flaming excuse, like this one.)

If you dwell on your objections, listing them and showing how each one is valid, your audience will be impressed more by their weight of numbers than by their lack of substance.

*I too like the idea of extending choice by having vending machines in trains, but there are eight objections. First, how would passengers manage to get the right coins for them? Second . . .*

(Very good, so long as you never mention the real objection, which is that they would enable people to bypass failures in the existing service. Stick to the trivia; it's safer ground.)

# *Tu quoque*

*Tu quoque* means 'you also', and is committed when a case is undermined by the claim that its proponent is himself guilty of what he talks of. It is a change of subject from a claim made by a proponent to one made against him. ('You accuse me of abusing my position, but you're the one whose company car is seen propping up the rails at the local race-course!')

With a little more subtlety, the *tu quoque* can be used to undermine an accusation by discrediting the accuser.

*And now I turn to Mrs Green's charge that I deliberately misled this society over my personal interest in the firm concerned. May I remind you that this charge comes from the same Mrs Green who kept very quiet when her son-in-law benefited from our decision over the surplus land. Hardly a source entitled to make such charges, you must agree.*

(I reckon he did it.)

The fallacy of the *tu quoque* occurs because it makes no attempt to deal with the subject under discussion. A new subject is introduced, namely the record of someone involved. The truth or falsehood of an assertion has nothing to do with the background of the person who makes it. Evidence for or against that assertion is not altered by details of the previous actions of the one who is putting it forward.

Another version of the *tu quoque* seeks to undermine what is being said by showing it to be inconsistent with the previous views of its proposer.

*Why should we listen to Brown's support for the new car park when only last year he opposed the whole idea?*

(For one thing, if the arguments changed his mind they might be worth listening to. For another, there might be more cars around.)

Because someone once opposed an idea it does not preclude their arguments in favour from being good ones. Despite this, the fallacy is supported by a strong tendency in us to appear consistent whenever we can. The new mayor finds it difficult to argue with sincerity this year in favour of the same official limousine which he opposed so vociferously for his predecessor.

The UK's parliamentary question-time is the home of the *tu quoque*. Indeed, skill at handling questions is often measured exclusively in terms of the performer's dexterity with this particular fallacy. This is why answers to questions about the present or the future invariably begin with the phrase:

*May I remind my honourable colleagues . . .*

(He is, of course, reminding those opposite that they did it sooner, longer, deeper, louder and worse. This is why their specious charges can be rejected.)

A parliamentary question is always known in the House as a 'PQ'. There is a good case to be made for having the reply to one of them called a 'TQ'.

The *tu quoque* is easy to use because everyone is inconsistent some of the time, and few people have a blameless past. You can argue that anyone who has changed their mind has thereby proved that they must be wrong at least some of the time, and that this occasion could well be one of those times. If you can find nothing at all to your opponent's discredit, even this fact can be used in an attempt to undermine what he is saying. The rest of us have weaknesses, why doesn't he?

*As for the charges that I may just occasionally have helped myself out of difficulty to a small extent, all I can say is look at Mr High-and-mighty Holier-than-thou.*

(And he is probably quite a lot holier than thou.)

# Unaccepted enthymemes

An enthymeme is an argument with one of its stages understood rather than stated. This is all right as long as both parties accept the tacit assumption. When the unstated element is not accepted, we move into the territory of the fallacy.

*Bill must be stupid. You have to be stupid to fail a driving-test.*

(While the average listener might nod sagely at this point, he would be somewhat put out if he later discovered that Bill hadn't failed his driving test. The argument only works if that is assumed.)

In this case a fallacy is committed because an important element of the argument is omitted. If both parties agree on the assumption, then it is present although unstated. If only the listener makes the assumption, he may think the argument has more support that it really does. We often leave out important stages because they are generally understood, but we have to recognize that there can be disagreements about what we are entitled to assume.

*I hope to repay the bank soon, Mr Smith. My late aunt said she would leave a reward to everyone who had looked after her.*

(The bank manager, surprised by the non-payment of the debt, will be even more surprised when you tell him how you had always neglected your aunt.)

It is because we use enthymemes routinely to avoid laboriously filling in the details that opportunities for the fallacy arise. The earnest caller who wishes to discuss the Bible with you will be satisfied if told 'I'm a Buddhist', because both parties accept the implicit fact that Buddhists do not discuss the Bible. If, however, you were to reply instead: 'Buddhists don't discuss the Bible', your caller might still be satisfied, making the obvious assumption that you were a Buddhist. (Make sure though that you have a very good answer ready should you happen to meet him in church next Sunday.)

Unaccepted enthymemes form ready crutches for lame excuses. The listener will generously clothe them with the unstated part necessary to complete the argument, instead of leaving them to blush naked.

*Darling, I'm sorry. Busy people tend to forget such things as anni-versaries.*

(This is fine until your colleagues mention that you've done nothing for two months except the *Telegraph* crossword.)

The fallacy is easy to use, and will get you off the hook in a wide variety of situations. The procedure is simple. Give a general statement as the answer to an individual situation. Your audience will automatically assume the missing premise: that the general situation applies to this particular case. What people normally do in certain circumstances is only relevant to the charges against you if it is assumed that you were indeed in those circumstances. The unaccepted enthymeme will slide in as smoothly as vintage port.

> *Yes, I am rather late. One simply cannot depend on buses and trains any more.*

> (True, but you walked from just around the corner.)

You can equally well make general assertions during a discussion about someone in particular. Your audience's delight at gossip and determination to believe the worst in everyone will help the unaccepted enthymeme to mingle with the invited guests.

> *I'm not happy with the choice of Smith. One can never be happy with those who prey on rich widows.*

> (Or on unjustified implications.)

# The undistributed middle

Classic among schoolboy fallacies is the argument that because all horses have four legs and all dogs have four legs, so all horses are dogs. This is the simplest version of the notorious fallacy of the undistributed middle. Both horses and dogs are indeed four-legged, but neither of them occupies the whole of the class of four-legged beings. This leaves convenient room for horses and dogs to be different from each other, and from other beings which might also without any overlap be in the four-legged class.

The 'middle' which carelessly omitted to get itself distributed is the term which appears in the first two lines of a three-line argument, but which disappears in the conclusion. The classic three-liner requires that this middle term must cover the whole of its class at least once. If not, it is undistributed.

*All men are mammals. Some mammals are rabbits, therefore some men are rabbits.*

(Even though the first two lines are correct, the middle term 'mammals' never once refers to *all* mammals. The middle term is thus undistributed and the deduction invalid.)

Common sense shows why the undistributed middle is fallacious. The standard three-liner (called a 'syllogism') works by relating one thing to another by means of a relationship they both have with a third. Only if at least one of those relationships applies to *all* the third thing do we know that it is certain to include the other relationship.

We cannot say that bureaucrats are petty tyrants just because bureaucrats are meddlesome and petty tyrants are meddlesome. It is quite possible that gin-sodden drunks are meddlesome too, but that does not mean that bureaucrats are gin-sodden drunks. (Life might be more interesting if they were.) This fallacy commonly appears in the form of 'tarring with the same brush.'

*The worst oppressors of the working class are landlords. Jones is a landlord, so Jones is one of the worst oppressors of the working classes.*

(Exit Jones, hurriedly, before it is pointed out that the worst oppressors of the working classes are human. Since Jones is human . . .)

The great thing about undistributed middles is that you can undistribute new ones as further 'evidence' in support of your previously undistributed ones. (The worst oppressors of the working class wear shoes; Jones wears shoes . . .)

The expert user will take the trouble to find out which terms are distributed or undistributed. He will learn the simple rule: *'Universals have distributed subjects, negatives have distributed predicates.'* Universals are statements which tell us about all or none of a class, and negatives tell us what isn't so. Armed with this technical information, the expert is able to inflict upon his audience such monstrosities as:

*All nurses are really great people, but it happens that some really great people are not properly rewarded. So some nurses are not properly rewarded.*

(It may be true, but has he given an argument? Since the middle term 'really great people' is neither the subject of a universal, nor the predicate of a negative, it is not distributed. We have here, therefore, a very complex fallacy of the undistributed middle.)

Leaving aside these technical uses, the fallacy in its simple form will give hours of pleasurable success if applied systematically. You should use it to gain approval for what you favour by pointing out how it shares qualities with something universally admired. Similarly, opposing ideas can be discredited by showing what qualities they share with universally detested things.

*The union closed shop is the will of the majority; and democracy is the will of the majority. The union closed shop is only democratic.*

(Where do I sign? [You did.])

*Elitism is something only a few benefit from, and tennis is something only a few benefit from, so tennis is clearly elitist.*

(Fault!)

# Unobtainable perfection

When the arguments for and against courses of action are assessed, it is important to remember that the choice has to be made from the available alternatives. All of them might be criticized for their imperfections, as might the status quo. Unless one of the options is perfect, the imperfections of the others are insufficient grounds for rejection. The fallacy of unobtainable perfection is committed when lack of perfection is urged as a basis for rejection, even though none of the alternatives is perfect either.

*We should ban the generation of nuclear power because it can never be made completely safe.*

(Also coal, oil and hydro-electric, all of which kill people every year in production and use. The question should be whether nuclear power would be better or worse than they are.)

If none of the alternatives, including making no change at all, is perfect, then imperfection is not grounds for a decision between them. To the matter of that choice it is irrelevant. If used to criticize only one option, it unfairly loads the case against that choice because it could be applied to all of them.

*I'm against going to the Greek islands because we cannot guarantee we would enjoy ourselves there.*

(When you do find a place for which this is guaranteed, let me know.)

The fallacy is very often used to reject changes to the status quo, even though the status quo itself might not be perfect.

*We must ban the new heart drug because it has been occasionally associated with neurological disorders.*

(This looks all right, but what if there are presently 15,000 patients dying each year of heart disease who could be saved by the new drug? Neither is the status quo perfect.)

Television documentaries and public affairs programmes are excellent source material for the unobtainable perfection fallacy. Any new proposal of government, any government, will be subjected to detailed analysis of its imperfections. Frail widows and struggling mothers will relate to cameras the hardships which will be caused, and the audience will be left with the uneasy feeling that the government is being too hasty. Exactly the same treatment could be given to the present situation.

The fallacy haunts the polished halls of committee meetings. On every committee is one person, usually a long-serving member, whose mission in life is to hold back the tide of anarchy and destruction which change represents. He castigates every new proposal with its own imperfections.

*I don't think banning cars from Park Street will prevent old people being hurt. There will still be children on roller-blades and bicycles, and shopping trolleys and baby carriages.*

(The question is not 'is it perfect?' The issue is whether the new proposal will cut down accidents as the status quo cuts down old people.)

While you can use the general version of this fallacy to undermine any proposals you disapprove of, it will also repay you if you take the time and trouble to learn two specialist and very clever versions of it. The first of these calls for a particular suggestion to be opposed because it does not go far enough. You show its imperfections, and suggest that something more drastic is needed. This idea, therefore, should be rejected.

*I approve in principle of the proposal to have the benefits allocated by lot, rather than by my personal decision, but this will still leave many areas of patronage and influence untouched. I suggest that a much wider measure is needed, looking at the whole field, and therefore I propose that we refer this suggestion back . . .*

(It was never seen again.)

The second variant you can use has you calling for something totally beyond the powers of those making the decision, and thus sets something they cannot do in opposition to something they can.

*It's all very well to suggest stiffer penalties for cheating, Headmaster, but that will not eradicate the problem. What we need instead is to win over these boys and girls, to effect a change in their hearts and minds . . .*

(The original proposal now exits amid a crescendo of violins.)

# *Verecundiam, argumentum ad*

This is the appeal to false authority. While it is perfectly in order to cite as a supporting witness someone who has specialized knowledge of the field concerned, it is a fallacy to suppose that an expert in one field can lend support in another. Unless he has special expertise, he is a false authority.

*Hundreds of leading scientists reject evolution.*

(Close examination shows few, if any, whose expertise is in evolutionary biology.)

Knowledge is specialized, and we have to accept the view of authorities to some extent. There is a general reluctance to challenge the view of someone who appears much more qualified than ordinary people. When support for a position is urged on account of the opinion of someone who appears to be more qualified but is not, the fallacy of *argumentum ad verecundiam* is committed.

The fallacy lies in the introduction of material that has no bearing on the matter under discussion. We have no reason to suppose that the opinion of a qualified person is worth any more than our own. The attempt to make our own opinions yield before such spurious authority is trading on our respect for position and achievement, and trying to use this instead of argument and evidence.

*The cologne of the stars.*

(Since few of us are lucky enough actually to smell our heroes and heroines, their opinions on this subject are probably less interesting than those of ordinary people closer at hand.)

The *argumentum ad verecundiam* dominates the world of advertising. Those who are thought worthy of admiration and esteem because of their achievements frequently descend to our level to give advice on more humdrum matters. Those whose excellence is in acting are only too ready to share with us their vast experience of instant coffee and dog-food. The winning of an Oscar for excellence in motion pictures is widely recognized as a qualification to speak on such matters as world poverty and American foreign policy.

One can admit the current young hopeful some authority on tennis rackets after a Wimbledon success; but razor blades? (One is surprised to find that he shaves.) In a similar way we see famous faces eating yoghurt or buying life assurance. Those who have proved their worth as presenters of radio or television programmes readily share with us their detailed expertise on enzyme-action washing powders or the virtues of a margarine which is high in polyunsaturates.

A variant of the *argumentum ad verecundiam* has the appeal to unidentified authorities, albeit those in the right field. In this world we are confronted by the opinions of 'leading scientists', 'top dog-breeders' and 'choosy mums'. Since we do not know who they are, all we can do is to accept the apparent authority they have. We never hear from the mediocre scientists, the average-to-poor dog-breeders or the indifferent mothers.

There is also the visual *ad verecundiam,* instanced by the sports team wearing the sponsor's name or slogan, even if unconnected with the sport.

*Winning the world slalom championships gives me a real thirst. That's why . . .*

(And the logic is as frothy as the stuff he's selling.)

Your own use of the *ad verecundiam* is made easier by the desire of many eminent people to be thought of as compassionate people with wide-ranging concerns. No matter how dotty the cause, you will always be able to assemble a panel of distinguished names to act as honorary patrons to it. The fact that they have achieved eminence as actors, writers and singing stars will in no ways diminish their authority to lend weight to your campaign.

*In demanding a ban on Spanish imports until bullfighting is out-lawed, I am joined by distinguished international scientists, top scholars and leading figures from the worlds of communication and the arts.*

(They should know. After all, they are also experts on wars, whales and windmills.)

# We must do something

People have a passion to right wrongs and to change the world for the better. This laudable attitude gave us vacuum cleaners and washing machines to right the wrongs of domestic chores. Sometimes, though, there seems to be little we can do that will make a difference.

**A** *We must protest against the civil war in Central Africa by having a 10,000 strong march through central London.*

**B** *But what impact will a London protest have on the combatants?*

**A** *Well we must do something.*

(Why not try standing on one leg and hopping up and down on the spot? It will be about as effective as a London protest march, and will be much easier to organize.)

People have used the notion that 'we must do something' to justify all manner of aimless marches, boycotts, trade embargoes and a refusal to wear items of clothing that started life being worn by animals. When people feel impelled to 'do something', no matter how small its impact on what they are trying to change, their actions are more about making themselves feel good than about altering reality. The fallacy consists in supposing that doing something ineffective is better than doing nothing at all. Not so. If the action is ineffective it is the same as doing nothing at all, but less easy to do.

**A** *I'm wearing a wristband to oppose forced marriages in South Asia.*

**B** *What difference will that make?*

**C** *I have to do something.*

(Wilberforce did more to combat slavery than wearing a wristband)

You can use the 'have to do something' approach yourself, justifying apparently random actions you undertake by linking them to worthwhile causes. Don't worry if the link is tenuous; most of them are. As long as you appear earnest and concerned, people will not mind if your actions are ineffective.

# Why wasn't it done already?

A joke which enjoys a certain popularity among economists asks how many free marketeers does it take to change a light bulb. The answer is zero because 'If the light bulb needed to be changed, the market would already have done that.' Leaving aside the humour, there is a fallacy in supposing that if a proposed idea is worthwhile, it would already have been done.

*If it's practical to drag suitcases along on wheels, how come it took over a hundred years before anyone did it?*

The fallacy occurs because there has to be a first time for everything. Every new invention, idea or proposal is at one stage the first of its kind. Maybe it never occurred to people that wheels on suitcases would make them easier to take through airports and train stations? Maybe someone finally became fed up of suffering hernias from heaving heavy bags around? Maybe someone watched a hotel porter wheel bags through a lobby on a trolley and thought they could fit the trolley to the bag?

There has to be a first time for everything, and it is clearly a fallacy to oppose an idea on the grounds that everything worthwhile has already been done. It is quite a good fallacy because of its plausibility. Some good ideas seem so blindingly obvious once uttered that it can seem extraordinary that no-one thought of it before.

Some people commit the fallacy because they sense that most new ideas are wacky, and that very few are worthwhile. This gives them a dismissive skepticism of new proposals, and they easily slip into the error of throwing out the few babies along with the positive torrent of bathwater.

*'If e-cigarettes are such a great idea, how come no-one thought of them before?'*

(Maybe people did, only to be ridiculed on the grounds that if the idea were any good it would already have been done.)

If you prefer things to be the way they have always been, this is your fallacy. When someone proposes an idea they say will improve things, just tell that person scornfully that it would be remarkable if they, alone out of billions of people before them, had produced an idea that the others just hadn't thought of. A tinge of sarcasm makes this more effective.

# Wishful thinking

While many of us engage quite happily in wishful thinking, we elevate it to the status of a fallacy when we use it in place of argument. If we accept a contention because we would like it to be true, rather than because of the arguments or evidence which support it, we move into fallacy. Similarly, we also commit the fallacy of wishful thinking if we reject something solely because we do not wish it to be true.

*Going to work in this awful weather would do no good for anyone. I think I'll take the day off and stay in bed.*

(Everyone must have felt the force of this argument at some time. Unfortunately, while there may be reasons for and against going into work, not wanting to is one which lacks persuasive force over everyone except ourselves.)

Our wishes rarely bear directly on the question of whether a thing is true or false. We commit a fallacy by intruding them into a discussion of the pros and cons. To suppose that the world is as we would want it to be is good solipsism but bad logic.

*Of course the environment talks will succeed. Otherwise it means mankind is on the way out.*

(The fact that we want them to succeed does not mean that they will. It could be that mankind *is* on the way out; in which case you might just as well be packing as hoping.)

Wishful thinking often appears to colour our judgement of outcomes we are unable to influence.

*He can't die. We couldn't manage without him.*

(He did. They could.)

Death, in fact, is a subject especially prone to the fallacy of wishful thinking. Its abrupt and inconsiderate nature is softened by the fallacy into something we would find more acceptable, although our wishes hardly afford valid grounds for our supposition. Boswell, on a visit to the dying Hume, asked the philosopher about a possible afterlife:

*Would it not be agreeable to have hopes of seeing our friends again?*

(He mentioned three recently deceased friends of Hume, but the latter firmly rejected the fallacy. 'He owned it would be agreeable', Boswell reported, 'but added that none of them entertained so absurd a notion.')

Time, like death, is a field in which our wishes replace our ability to influence.

*It can't be Friday already! I've not done nearly enough work to pass the exam!*

(Wrong about the day; right about the exam.)

The problem about all wishful thinking is that if you want one thing and the laws of the universe dictate another there is a conflict of interests which is not going to be resolved in your favour. This being true, you might as well spend time working out how to deal with the outcome, instead of wishing that something else would happen.

*The bank will extend our overdraft; otherwise we just cannot survive.*

(Bank managers are not interested in your survival. They care about only two things: making money for the bank and grinding the faces of the poor.)

Most of us are already fairly adept at using the fallacy of wishful thinking to persuade ourselves. When using it to convince others, bear in mind that it must be their wishes, rather than your own, which are appealed to.

*The business will succeed. You'll get a huge return on your investment.*

(This is more effective than 'The business will succeed. I'll be rich for life!')

# CLASSIFICATION OF FALLACIES

There are five broad categories into which fallacies fall. The most important division is between the formal fallacies and the informal ones, although there are important distinctions between the various types of informal fallacy.

Formal fallacies have some error in the structure of the logic. Although they often resemble valid forms of argument, the staircase only takes us from A to B by way of cracked or missing steps. In brief, the fallacy occurs because the chain of reasoning itself is defective.

Informal fallacies, on the other hand, often use valid reasoning on terms which are not of sufficient quality to merit such treatment. They can be linguistic, allowing ambiguities of language to admit error; or they can be fallacies of relevance which omit something needed to sustain the argument, permit irrelevant factors to weigh on the conclusion, or allow unwarranted presumptions to alter the conclusion reached.

The five categories of fallacy are:

1   formal

2   informal (linguistic)

3   informal (relevance – omission)

4   informal (relevance – intrusion)

5   informal (relevance – presumption)

# The formal fallacies

Affirming the consequent
Conclusion which denies premises
Contradictory premises
Denying the antecedent
Exclusive premises
Existential fallacy
False conversion
Illicit process (minor)
Illicit process (major)
Positive conclusion/negative premises
*Quaternio terminorum*
Undistributed middle

# The informal linguistic fallacies

Accent
Amphiboly
Composition
Division
Equivocation
Reification

# The informal fallacies of relevance (omission)

Bogus dilemma
Concealed quantification
Damning the alternatives

Definitional retreat
Extensional pruning
Hedging
*Ignorantiam, argumentum ad*
Insulation from alternatives
It's worth it if it saves lives
*Lapidem, argumentum ad*
*Nauseam, argumentum ad*
Omitting the benefit
One-sided assessment
Refuting the example
Shifting ground
Shifting the burden of proof
Special pleading
Straw man
The exception that proves the rule
Trivial objections
Unaccepted enthymemes
Unobtainable perfection

# The informal fallacies of relevance (intrusion)

Anecdotal argument
Blinding with science
Buzzwords
Collective guilt
*Crumenam, argumentum ad*
Emotional appeals
    *(argumentum ad invidiam)*
    *(argumentum ad metum)*
    *(argumentum ad modum)*

*(argumentum ad odium)*
*(argumentum ad superbiam)*
*(argumentum ad superstitionem)*
*(sentimens superior)*
Every schoolboy knows
Genetic fallacy
*Hominem* (abusive) *argumentum ad*
*Hominem* (circumstantial) *argumentum ad*
*Ignoratio elenchi*
Irrelevant humour
*Lazarum, argumentum ad*
Loaded words
*Misericordiam, argumentum ad*
Poisoning the well
*Populum, argumentum ad*
Presentation over content
The red herring
The runaway train
The slippery slope
*Tu quoque*
*Verecundiam, argumentum ad*
We must do something
Wishful thinking

# The informal fallacies of relevance (presumption)

Abusive analogy
Accident
Analogical fallacy
*Antiquitam, argumentum ad*
Apriorism

Bifurcation
*Circulus in probando*
Complex questions
*Cum hoc ergo propter hoc*
In denial
*Dicto simpliciter*
Ethical superiority
*Ex-post-facto* statistics
False zero sum game
The gambler's fallacy
Non-anticipation
*Novitam, argumentum ad*
*Petitio principii*
*Post hoc ergo propter hoc*
*Secundum quid*
*Temperantiam, argumentum ad*
Thatcher's blame
Why wasn't it done already